OXFORD MATHEMATICAL MONOGRAPHS

Editors

G. TEMPLE   I. JAMES

OXFORD MATHEMATICAL MONOGRAPHS

COMPLEMENTARY VARIATIONAL PRINCIPLES

BY

A. M. ARTHURS

CLARENDON PRESS · OXFORD

1970

Oxford University Press, Ely House, London W.1

GLASGOW   NEW YORK   TORONTO   MELBOURNE   WELLINGTON
CAPE TOWN   SALISBURY   IBADAN   NAIROBI   DAR ES SALAAM   LUSAKA   ADDIS ABABA
BOMBAY   CALCUTTA   MADRAS   KARACHI   LAHORE   DACCA
KUALA LUMPUR   SINGAPORE   HONG KONG   TOKYO

PRINTED IN NORTHERN IRELAND

AT THE UNIVERSITIES PRESS, BELFAST

TO

MY MOTHER

PREFACE

ONE of the earliest examples of complementary variational principles is provided by the energy principle of Dirichlet in the theory of electrostatics, together with the Thomson principle of complementary energy. Another example is that given by the Ritz and Temple bounds for eigenvalues. There are several methods by which maximum and minimum (complementary) variational principles can be derived. One such method, due to Friedrichs, involves the use of involutory transformations, while another, which applies to certain linear problems, employs the hypercircle approach of Prager and Synge.

However, one of the most straightforward methods of deriving complementary principles is based on the canonical form of the Euler–Lagrange variational theory. This method, which forms the subject of this monograph, was introduced by Noble in 1964, and has provided a systematic approach to many linear and nonlinear boundary-value problems involving differential, integral, and matrix equations.

Chapter I provides a simple introduction to the basic ideas of complementary variational principles and deals with the theory for the simplest problem in the calculus of variations.

A natural extension of the theory to cover a certain class of operator equations and associated boundary-value problems forms the subject of Chapter 2.

The remainder of the book, comprising Chapters 3 and 4, is concerned with applications of the theory to particular kinds of linear and nonlinear boundary-value problems. The associated complementary variational principles are illustrated by various examples taken from mathematical physics. These examples range over potential theory, transport theory, quantum perturbation and scattering theory, hydrodynamics, networks, and communication theory. As well as containing new derivations of well-known results such as the Ritz and Temple bounds for eigenvalues, the examples contain many results on upper and lower bounds that have only recently been obtained.

The book is written at a fairly elementary level, and should be accessible to any student with a little knowledge of calculus of variations and linear operator theory.

I wish to thank Professor C. A. Coulson of Oxford, Professor B. Noble of Wisconsin, and Dr. N. Anderson and Dr. P. D. Robinson of York, for many helpful discussions and suggestions on the material presented here. I am also grateful to the editors and the Clarendon Press for including this monograph in their series.

A. M. A.

University of York
September, 1969

CONTENTS

1

VARIATIONAL PRINCIPLES: INTRODUCTION

1.1. Introduction

VARIATIONAL principles play an important part in mathematics and the physical sciences for three main reasons: they (i) unify many diverse fields, (ii) lead to new theoretical results, and (iii) provide powerful methods of calculation. Thus, the well-known Euler–Lagrange principle can be used to derive field equations of many kinds, extremum principles lead to new estimates for important physical quantities, and direct methods form the basis of very accurate computations [cf. 36, 52, 54]. Many problems, however, are usually first posed in the form of differential equations, or more generally as operator equations, and there is no guarantee that an equivalent variational problem exists. Even if we know that the equivalent problem does exist, it may not be easy to find an explicit form for the variational expression. Stated in mathematical terms, the problem is to find the potential corresponding to a given field equation [cf. 82]. Of course in some branches of mathematical physics, such as analytical dynamics, the variational problem is known once the Lagrangian is specified. As it turns out, all the results obtained in this book are examples of this latter kind, for which the potential or basic functional is easily found. Our particular interest centres on principles which lead to variational bounds, and in particular those cases for which both maximum and minimum (complementary) variational principles can be obtained. In many applications these complementary variational principles provide upper and lower bounds for quantities of interest.

One of the earliest examples of complementary principles is provided by the energy principle in the theory of structures, together with the principle of complementary energy [80]. Another example concerns the Dirichlet and Thomson bounds in electrostatics. There are several methods by which complementary principles can be derived. One such method involves the use of involutory transformations [28], while another, which applies to certain linear problems, employs the hypercircle approach [78].

However, one of the most straightforward methods of deriving complementary principles is based on the canonical form of the Euler–Lagrange variational theory. This method, which forms the subject of the present monograph, was introduced by Noble [55] in 1964, and has provided a systematic approach to many linear and nonlinear problems involving differential, integral, and matrix equations.

1.2. Euler–Lagrange theory

The variational principles described in this book have their origins in the simplest kind of variational problem which can be treated by the Euler–Lagrange theory. Thus they are basically concerned with differentiable functionals of the form

$$E(\Phi) = \int_a^b L(x, \Phi, \Phi')\, dx, \qquad \Phi' = d\Phi/dx, \tag{1.2.1}$$

with fixed end-points

$$\Phi(a) = \alpha, \qquad \Phi(b) = \beta. \tag{1.2.2}$$

Here Φ belongs to the class C_2 of functions which have continuous derivatives up to second order for $a \leqslant x \leqslant b$, and L is assumed to possess continuous second-order derivatives with respect to all its arguments. Of course, the assumptions just made can be relaxed to some considerable extent [cf. 35, 59], but we shall not deal with that aspect of the theory here.

Suppose now that the functional $E(\Phi)$ has an extremum at φ. Then we consider variations round φ

$$\Phi = \varphi + \epsilon\xi. \tag{1.2.3}$$

If Φ and φ both satisfy the end-point conditions (1.2.2), it follows that

$$\xi(a) = \xi(b) = 0. \tag{1.2.4}$$

Since $E(\Phi)$ is differentiable, we can write

$$E(\varphi + \epsilon\xi) = E(\varphi) + \delta E(\varphi, \epsilon\xi) + \delta^2 E(\varphi, \epsilon\xi) + \ldots, \tag{1.2.5}$$

where the first variation is

$$\delta E = \epsilon \int_a^b \left\{ \xi\left(\frac{\partial L}{\partial \Phi}\right)_\varphi + \xi'\left(\frac{\partial L}{\partial \Phi'}\right)_\varphi \right\} dx \tag{1.2.6}$$

and the second variation is

$$\delta^2 E = \tfrac{1}{2}\epsilon^2 \int_a^b \left\{ \xi\left(\frac{\partial^2 L}{\partial \Phi^2}\right)_\varphi \xi + 2\xi\left(\frac{\partial^2 L}{\partial \Phi\,\partial \Phi'}\right)_\varphi \xi' + \xi'\left(\frac{\partial^2 L}{\partial \Phi'^2}\right)_\varphi \xi' \right\} dx. \quad (1.2.7)$$

Integrating by parts in (1.2.6), we obtain an alternative expression for the first variation

$$\delta E = \epsilon \int_a^b \xi \left\{\frac{\partial L}{\partial \Phi} - \frac{\mathrm{d}}{\mathrm{d}x}\left(\frac{\partial L}{\partial \Phi'}\right)\right\}_\varphi \mathrm{d}x + \left[\xi\left(\frac{\partial L}{\partial \Phi'}\right)_\varphi\right]_a^b. \quad (1.2.8)$$

Since the variations are such that ξ vanishes at the end-points, this reduces to

$$\delta E = \epsilon \int_a^b \xi \left\{\frac{\partial L}{\partial \Phi} - \frac{\mathrm{d}}{\mathrm{d}x}\left(\frac{\partial L}{\partial \Phi'}\right)\right\}_\varphi \mathrm{d}x. \quad (1.2.9)$$

For the functional $E(\Phi)$ to have an extremum, that is, be stationary at $\Phi = \varphi$, it is necessary that the first variation vanish. From (1.2.9) this means that

$$\int_a^b \xi \left\{\frac{\partial L}{\partial \Phi} - \frac{\mathrm{d}}{\mathrm{d}x}\left(\frac{\partial L}{\partial \Phi'}\right)\right\}_\varphi \mathrm{d}x = 0. \quad (1.2.10)$$

Since ξ is arbitrary in the interval (a, b), it follows from (1.2.10) that

$$\left\{\frac{\partial L}{\partial \Phi} - \frac{\mathrm{d}}{\mathrm{d}x}\left(\frac{\partial L}{\partial \Phi'}\right)\right\}_\varphi = 0, \qquad a \leqslant x \leqslant b, \quad (1.2.11)$$

which is the Euler–Lagrange equation. We therefore can state

THEOREM 1.2.1. $E(\Phi)$ is stationary at φ where φ is the solution of

$$\frac{\partial L}{\partial \Phi} - \frac{\mathrm{d}}{\mathrm{d}x}\left(\frac{\partial L}{\partial \Phi'}\right) = 0, \qquad a < x < b \quad (1.2.12)$$

with
$$\Phi(a) = \alpha, \qquad \Phi(b) = \beta. \quad (1.2.13)$$

This is the Euler–Lagrange variational principle.

Equation (1.2.10) gives a necessary condition for an extremum, but, in general, one which is not sufficient. In many cases, however, the Euler–Lagrange equation (1.2.12) by itself is enough to give a complete solution of the problem, and in fact the existence of an extremum is often clear from the physical meaning of the problem. If in such a case there exists only one extremal φ satisfying the boundary conditions of the problem, this extremal must be the function for which the extremum is attained.

Assuming that we have found the function φ which makes $E(\Phi)$ stationary, we now wish to consider the nature of the extremum, that is, its maximum, minimum, or saddle-point properties. To do this we shall look at the second variation defined in (1.2.7). If terms of $O(\epsilon^3)$ can be neglected in (1.2.5), or if they vanish as is the case for quadratic L, it follows from (1.2.5) that

THEOREM 1.2.2. *A necessary condition for the functional $E(\Phi)$ to have a minimum, $E(\Phi) \geqslant E(\varphi)$, for $\Phi = \varphi$, is that*

$$\delta^2 E(\varphi, \epsilon\xi) \geqslant 0 \tag{1.2.14}$$

for all admissible ξ, where $\delta^2 E$ is given by (1.2.7). Similarly, for a maximum, $E(\Phi) \leqslant E(\varphi)$, at $\Phi = \varphi$, the condition is

$$\delta^2 E(\varphi, \epsilon\xi) \leqslant 0. \tag{1.2.15}$$

The two cases in this theorem enable us therefore to obtain upper or lower bounds for the stationary value $E(\varphi)$ of the functional.

Example. To illustrate these results we consider a quadratic L given by

$$L = \tfrac{1}{2}v(\Phi')^2 + \tfrac{1}{2}w\Phi^2 - q\Phi, \tag{1.2.16}$$

where v, w, and q may be functions of x, with $v > 0$, $w \geqslant 0$. The variational problem associated with (1.2.1) is then equivalent to the boundary-value problem

$$-\frac{d}{dx}\left\{v\,\frac{d\varphi}{dx}\right\} + w\varphi = q, \qquad a \leqslant x \leqslant b, \tag{1.2.17}$$

$$\varphi(a) = \alpha, \qquad \varphi(b) = \beta, \tag{1.2.18}$$

involving a Sturm–Liouville equation (1.2.17). These equations follow directly by setting (1.2.16) in (1.2.12). From (1.2.5), (1.2.7), and (1.2.16) we find that the basic functional

$$E(\Phi) = \tfrac{1}{2}\int_a^b \{v(\Phi')^2 + w\Phi^2 - 2q\Phi\}\,dx \tag{1.2.19}$$

may be expanded as

$$E(\Phi) = E(\varphi) + \delta^2 E(\varphi, \epsilon\xi), \tag{1.2.20}$$

where
$$\delta^2 E(\varphi, \epsilon\xi) = \tfrac{1}{2}\epsilon^2 \int_a^b \{w\xi^2 + v(\xi')^2\}\,\mathrm{d}x, \qquad (1.2.21)$$

there being no third and higher order terms in (1.2.20) since L is quadratic here. Now $v > 0$ and $w \geqslant 0$, and so the second variation (1.2.21) is non-negative. Hence from (1.2.20) we can derive the minimum principle

$$E(\Phi) \geqslant E(\varphi). \qquad (1.2.22)$$

If we take $v = 1$, $w = q = 0$, this result corresponds to a one-dimensional form of the well-known Dirichlet principle [cf. 59].

These results show that variational problems which are formulated in terms of finding relative minima (or maxima) of Euler functionals lead in a natural way only to upper (or lower) bounds for the stationary value of the functional. However, certain minimization problems in the calculus of variations can be transformed into maximization problems. From a combination of these two problems it is then possible to obtain upper *and* lower bounds on the stationary value. One approach to these complementary bounds, due to Friedrichs [cf. 28], is based on involutory transformations. An alternative approach, introduced by Noble [55], is based on the canonical form of Euler theory, and it is this method which we now consider.

1.3. Canonical theory

The Euler–Lagrange principle discussed in section 1.2 can be readily generalized for functionals depending on two or more independent functions [cf. 35]. Thus the functions which make the Euler functional

$$E(U, \Phi) = \int_a^b L(x, U, \Phi, U', \Phi')\,\mathrm{d}x \qquad (1.3.1)$$

stationary, are given by $U = u$, $\Phi = \varphi$, where u and φ are the solutions of the pair of differential equations

$$\frac{\partial L}{\partial U} - \frac{\mathrm{d}}{\mathrm{d}x}\left(\frac{\partial L}{\partial U'}\right) = 0; \qquad (1.3.2)$$

$$\frac{\partial L}{\partial \Phi} - \frac{\mathrm{d}}{\mathrm{d}x}\left(\frac{\partial L}{\partial \Phi'}\right) = 0, \qquad (1.3.3)$$

and the trial functions U and Φ in (1.3.1) satisfy certain boundary conditions in general.

Now consider the case

$$L = U\frac{d\Phi}{dx} - H(x,\, U,\, \Phi). \tag{1.3.4}$$

For this choice of L the Euler–Lagrange equations (1.3.2) and (1.3.3) become

$$\frac{d\Phi}{dx} = \frac{\partial H}{\partial U}, \qquad -\frac{dU}{dx} = \frac{\partial H}{\partial \Phi}. \tag{1.3.5}$$

These are the so-called canonical Euler (or Hamilton) equations, which are well known for the part they play in analytical dynamics [46].

To discuss the variational theory associated with (1.3.5) in detail, we introduce the differentiable functional

$$I(U,\, \Phi) = \int_a^b \left\{ U\frac{d\Phi}{dx} - H(x,\, U,\, \Phi) \right\} dx + [\Gamma(U,\, \Phi) - U\Phi]_a^b, \tag{1.3.6}$$

where Γ is a boundary function determined by the particular form of boundary conditions that we require. Taking small variations round u and φ by setting

$$U = u + \epsilon\eta, \qquad \Phi = \varphi + \epsilon\xi \tag{1.3.7}$$

in (1.3.6), we obtain

$$I(U,\, \Phi) = I(u,\, \varphi) + \delta I + \delta^2 I + O(\epsilon^3), \tag{1.3.8}$$

where

$$\delta I = \epsilon \int_a^b \left\{ u\xi' + \eta\varphi' - \xi\left(\frac{\partial H}{\partial \Phi}\right)_{u,\varphi} - \eta\left(\frac{\partial H}{\partial U}\right)_{u,\varphi} \right\} dx +$$
$$+ \epsilon\left[\eta\left(\frac{\partial \Gamma}{\partial U}\right)_{u,\varphi} + \xi\left(\frac{\partial \Gamma}{\partial \Phi}\right)_{u,\varphi} - u\xi - \eta\varphi \right]_a^b, \tag{1.3.9}$$

and

$$\delta^2 I = \tfrac{1}{2}\epsilon^2 \int_a^b \left\{ 2\eta\xi' - \eta\left(\frac{\partial^2 H}{\partial U^2}\right)_{u,\varphi}\eta - 2\eta\left(\frac{\partial^2 H}{\partial U\, \partial \Phi}\right)_{u,\varphi}\xi - \xi\left(\frac{\partial^2 H}{\partial \Phi^2}\right)_{u,\varphi}\xi \right\} dx +$$
$$+ \tfrac{1}{2}\epsilon^2\left[\eta\left(\frac{\partial^2 \Gamma}{\partial U^2}\right)_{u,\varphi}\eta + 2\eta\left(\frac{\partial^2 \Gamma}{\partial U\, \partial \Phi}\right)_{u,\varphi}\xi + \xi\left(\frac{\partial^2 \Gamma}{\partial \Phi^2}\right)_{u,\varphi}\xi - 2\eta\xi \right]_a^b, \tag{1.3.10a}$$

or

$$\delta^2 I = \tfrac{1}{2}\epsilon^2 \int_a^b \left\{ 2\left(-\frac{d\eta}{dx}\right)\xi - \eta\left(\frac{\partial^2 H}{\partial U^2}\right)_{u,\varphi}\eta - 2\eta\left(\frac{\partial^2 H}{\partial U\, \partial \Phi}\right)_{u,\varphi}\xi - \xi\left(\frac{\partial^2 H}{\partial \Phi^2}\right)_{u,\varphi}\xi \right\} dx +$$
$$+ \tfrac{1}{2}\epsilon^2\left[\eta\left(\frac{\partial^2 \Gamma}{\partial U^2}\right)_{u,\varphi}\eta + 2\eta\left(\frac{\partial^2 \Gamma}{\partial U\, \partial \Phi}\right)_{u,\varphi}\xi + \xi\left(\frac{\partial^2 \Gamma}{\partial \Phi^2}\right)_{u,\varphi}\xi \right]_a^b, \tag{1.3.10b}$$

on integrating the first term of the previous line by parts.

For the functional (1.3.6) to be stationary at $U = u$, $\Phi = \varphi$, it is necessary that $\delta I = 0$. From (1.3.9) this means that

$$\int_a^b \left\{ \eta\left(\frac{d\Phi}{dx} - \frac{\partial H}{\partial U}\right)_{u,\varphi} + \xi\left(-\frac{dU}{dx} - \frac{\partial H}{\partial \Phi}\right)_{u,\varphi} \right\} dx +$$

$$+ \left[\eta\left(\frac{\partial \Gamma}{\partial U} - \Phi\right)_{u,\varphi} + \xi\left(\frac{\partial \Gamma}{\partial \Phi}\right)_{u,\varphi} \right]_a^b = 0, \quad (1.3.11)$$

where in (1.3.9) we have integrated the term involving $u\xi'$ by parts. We now choose the boundary function Γ to be

$$\Gamma = \begin{cases} U\alpha & \text{at} \quad x = a \\ U\beta & \text{at} \quad x = b, \end{cases} \quad (1.3.12)$$

and the functional $I(U, \Phi)$ in (1.3.6) becomes

$$I(U, \Phi) = \int_a^b \left\{ U\frac{d\Phi}{dx} - H(x, U, \Phi) \right\} dx + U(b)\{\beta - \Phi(b)\} -$$

$$- U(a)\{\alpha - \Phi(a)\}, \quad (1.3.13)$$

$$= \int_a^b \left\{ \left(-\frac{dU}{dx}\right)\Phi - H(x, U, \Phi) \right\} dx + \beta U(b) - \alpha U(a). \quad (1.3.14)$$

From (1.3.11) we therefore obtain the following principle for the functional $I(U, \Phi)$ defined by (1.3.13):

THEOREM 1.3.1. *The functional $I(U, \Phi)$ in (1.3.13) and (1.3.14) is stationary at u, φ, where u, φ are the solutions of the boundary-value problem*

$$\frac{d\Phi}{dx} = \frac{\partial H}{\partial U} \qquad a \leqslant x \leqslant b, \quad (1.3.15)$$

$$-\frac{dU}{dx} = \frac{\partial H}{\partial \Phi} \qquad a \leqslant x \leqslant b, \quad (1.3.16)$$

$$\Phi(a) = \alpha, \qquad \Phi(b) = \beta. \quad (1.3.17)$$

This result tells us the precise form of the boundary conditions associated with the functional (1.3.13). Other boundary conditions can also be included in the theory by choosing appropriate forms for Γ, but we shall concentrate here on conditions of the form (1.3.17).

Now we turn to the derivation of extremum principles, that is, maximum and minimum principles in which we say that under certain circumstances $I(u, \varphi)$ is greater than or less than $I(U, \Phi)$. Such results, however, are not possible in the general theory developed so far, where the trial functions U and Φ are arbitrary and independent. Some restriction on the trial functions is necessary, and involves making one trial function dependent on the other in a certain way [55].

Suppose first that we choose a trial function Φ, and then determine U as a function $Y(\Phi)$ of Φ by satisfying

$$\frac{d\Phi}{dx} = \frac{\partial H}{\partial U} \tag{1.3.15}$$

identically. This $U = Y$ is then substituted in (1.3.13), and we write the resulting functional as

$$I(Y(\Phi),\, \Phi) = J(\Phi). \tag{1.3.18}$$

It is clear from theorem 1.3.1 that the functional $J(\Phi)$ is stationary at φ and we write

$$J(\Phi) = I(u,\, \varphi) + \delta^2 J + O(\epsilon^3), \tag{1.3.19}$$

where from (1.3.10a) and (1.3.12)

$$\delta^2 J = \tfrac{1}{2}\epsilon^2 \int_a^b \left\{ 2\eta\xi' - \eta\left(\frac{\partial^2 H}{\partial U^2}\right)_{u,\varphi} \eta - 2\eta\left(\frac{\partial^2 H}{\partial U\,\partial\Phi}\right)_{u,\varphi} \xi - \xi\left(\frac{\partial^2 H}{\partial\Phi^2}\right)_{u,\varphi} \xi \right\} dx -$$
$$-\epsilon^2[\eta\xi]_a^b, \tag{1.3.20}$$

in which $\epsilon\xi = \Phi - \varphi$ and $\epsilon\eta = Y(\Phi) - u$. From (1.3.15) we find that

$$\frac{d\xi}{dx} = \left(\frac{\partial^2 H}{\partial U^2}\right)_{u,\varphi} \eta + \left(\frac{\partial^2 H}{\partial U\,\partial\Phi}\right)_{u,\varphi} \xi + O(\epsilon). \tag{1.3.21}$$

If, in addition, we choose Φ so that

$$\Phi(a) = \alpha, \qquad \Phi(b) = \beta, \tag{1.3.22}$$

giving

$$\xi(a) = 0, \qquad \xi(b) = 0, \tag{1.3.23}$$

equation (1.3.20) becomes

$$\delta^2 J = \frac{1}{2} \int_a^b \left\{ (Y(\Phi) - u)\left(\frac{\partial^2 H}{\partial U^2}\right)_{u,\varphi} (Y(\Phi) - u) - (\Phi - \varphi)\left(\frac{\partial^2 H}{\partial\Phi^2}\right)_{u,\varphi} (\Phi - \varphi) \right\} dx. \tag{1.3.24}$$

We collect these results together in

THEOREM 1.3.2. *The functional $J(\Phi)$ defined in (1.3.18) is stationary as Φ varies round φ, the exact solution of (1.3.15)–(1.3.17). In addition, if Φ satisfies the exact boundary conditions (1.3.17),*

$$J(\Phi) \equiv I(Y(\Phi),\, \Phi) = I(u,\, \varphi) + \delta^2 J + O(\epsilon^3), \tag{1.3.25}$$

where
$$\delta^2 J = \tfrac{1}{2}\epsilon^2 \int_a^b \left\{ \eta \left(\frac{\partial^2 H}{\partial U^2}\right)_{u,\varphi} \eta - \xi \left(\frac{\partial^2 H}{\partial \Phi^2}\right)_{u,\varphi} \xi \right\} dx, \tag{1.3.26}$$

with $\epsilon\eta = Y(\Phi) - u$, *and* $\epsilon\xi = \Phi - \varphi$. *If* $O(\epsilon^3)$ *terms are negligible in* (1.3.25), *it follows that*

$$J(\Phi) \leqslant I(u, \varphi) \quad if \quad \delta^2 J \leqslant 0, \tag{1.3.27a}$$
or $\quad\quad J(\Phi) \geqslant I(u, \varphi) \quad if \quad \delta^2 J \geqslant 0. \tag{1.3.27b}$

The complementary variational principle is obtained by considering the expression (1.3.14):

$$I(U, \Phi) = \int_a^b \left\{ \left(-\frac{dU}{dx}\right)\Phi - H(x, U, \Phi) \right\} dx + \beta U(b) - \alpha U(a). \tag{1.3.28}$$

In this case we first guess U and then assume that Φ is determined as a function $\Theta(U)$ of U in such a way that the equation

$$-\frac{dU}{dx} = \frac{\partial H}{\partial \Phi} \tag{1.3.16}$$

is satisfied identically. This $\Phi = \Theta(U)$ is then substituted in (1.3.28) and we write the resulting functional as

$$I(U, \Theta(U)) = G(U). \tag{1.3.29}$$

From theorem 1.3.1 it is clear that the functional $G(U)$ is stationary at u and we write
$$G(U) = I(u, \varphi) + \delta^2 G + O(\epsilon^3), \tag{1.3.30}$$
where from (1.3.10b) and (1.3.12)

$$\delta^2 G = \tfrac{1}{2}\epsilon^2 \int_a^b \left\{ 2\left(-\frac{d\eta}{dx}\right)\xi - \eta\left(\frac{\partial^2 H}{\partial U^2}\right)_{u,\varphi} \eta - 2\eta\left(\frac{\partial^2 H}{\partial U\,\partial \Phi}\right)_{u,\varphi} \xi - \xi\left(\frac{\partial^2 H}{\partial \Phi^2}\right)_{u,\varphi} \xi \right\} dx, \tag{1.3.31}$$

in which $\epsilon\xi = \Theta(U) - \varphi$ and $\epsilon\eta = U - u$. From (1.3.16) we find that

$$-\frac{d\eta}{dx} = \left(\frac{\partial^2 H}{\partial \Phi\,\partial U}\right)_{u,\varphi} \eta + \left(\frac{\partial^2 H}{\partial \Phi^2}\right)_{u,\varphi} \xi + O(\epsilon). \tag{1.3.32}$$

Hence (1.3.31) may be simplified to the form

$$\delta^2 G = -\frac{1}{2} \int_a^b \left\{ (U-u)\left(\frac{\partial^2 H}{\partial U^2}\right)_{u,\varphi} (U-u) - \right.$$
$$\left. - (\Theta(U)-\varphi)\left(\frac{\partial^2 H}{\partial \Phi^2}\right)_{u,\varphi} (\Theta(U)-\varphi) \right\} dx. \tag{1.3.33}$$

These results are now collected together in

THEOREM 1.3.3. *The functional* $G(U)$ *defined in* (1.3.29) *is stationary as* U *varies round* u, *the exact solution appearing in theorem* 1.3.1. *In addition*

$$G(U) \equiv I(U, \Theta(U)) = I(u, \varphi) + \delta^2 G + O(\epsilon^3), \qquad (1.3.34)$$

where

$$\delta^2 G = -\tfrac{1}{2}\epsilon^2 \int_a^b \left\{ \eta \left(\frac{\partial^2 H}{\partial U^2}\right)_{u,\varphi} \eta - \xi \left(\frac{\partial^2 H}{\partial \Phi^2}\right)_{u,\varphi} \xi \right\} dx, \qquad (1.3.35)$$

with $\epsilon\eta = U - u$, *and* $\epsilon\xi = \Theta(U) - \varphi$. *If* $O(\epsilon^3)$ *terms are negligible in* (1.3.34), *it follows that*

$$G(U) \leqslant I(u, \varphi) \qquad if \quad \delta^2 G \leqslant 0, \qquad (1.3.36a)$$

or

$$G(U) \geqslant I(u, \varphi) \qquad if \quad \delta^2 G \geqslant 0. \qquad (1.3.36b)$$

This principle is complementary to the one stated in theorem 1.3.2.

1.4. Complementary variational principles

We are now in a position to state the important result of Noble [55]. This follows directly from the extremum principles derived in theorems 1.3.2 and 1.3.3, and may be stated as

THEOREM 1.4.1. *With the assumptions of theorems* 1.3.2 *and* 1.3.3

$$G(U) \leqslant I(u, \varphi) \leqslant J(\Phi) \qquad if \quad \delta^2 G \leqslant 0, \delta^2 J \geqslant 0, \qquad (1.4.1)$$

equality holding when U *and* Φ *are the exact solutions. This result also holds if all the inequality signs are reversed.*

Formula (1.4.1) is the reason why the variational principles (1.3.19) and (1.3.30) are said to be complementary.

The maximum and minimum principles just derived give complementary upper and lower bounds for $I(u, \varphi)$, provided that $\delta^2 G$ and $\delta^2 J$ do not have the same sign, and provided that U and Φ are sufficiently close to u and φ. If $\delta^2 G$ and $\delta^2 J$ do have the same sign, the functionals $G(U)$ and $J(\Phi)$ become different one-sided bounds. The pair of functions (u, φ) furnishes the exact solution of the problem described by (1.3.15)–(1.3.17).

While the present analysis has been based on local variations, the behaviour with respect to large variations can in many cases be established without much extra difficulty [cf. 41, 75].

Example 1. These results on complementary variational principles may be illustrated by means of the boundary-value problem in equations (1.2.17) and (1.2.18) of the previous example. Suppose then that φ is the exact solution of

$$-\frac{\mathrm{d}}{\mathrm{d}x}\left\{v\,\frac{\mathrm{d}\Phi}{\mathrm{d}x}\right\}+w\Phi = q \qquad a \leqslant x \leqslant b, \tag{1.4.2}$$

$$\Phi(a) = \alpha, \qquad \Phi(b) = \beta, \tag{1.4.3}$$

where v, w, and q may be functions of x and we assume that $v > 0$, $w > 0$. To use the canonical form of the Euler–Lagrange equation we write (1.4.2) as the pair of equations

$$\frac{\mathrm{d}\Phi}{\mathrm{d}x} = \frac{1}{v}\,U = \frac{\partial H}{\partial U}, \tag{1.4.4}$$

$$-\frac{\mathrm{d}U}{\mathrm{d}x} = -w\Phi + q = \frac{\partial H}{\partial \Phi}. \tag{1.4.5}$$

A suitable H function is given by

$$H = \frac{1}{2v}\,U^2 - \tfrac{1}{2}w\Phi^2 + q\Phi. \tag{1.4.6}$$

From (1.3.13) and (1.3.14) we see that the basic functional $I(U, \Phi)$ in this example is given by

$$I(U, \Phi) = \int_a^b \left\{U\frac{\mathrm{d}\Phi}{\mathrm{d}x} - \frac{1}{2v}\,U^2 + \tfrac{1}{2}w\Phi^2 - q\Phi\right\}\mathrm{d}x +$$
$$+ U(b)\{\beta - \Phi(b)\} - U(a)\{\alpha - \Phi(a)\}, \tag{1.4.7}$$

$$= \int_a^b \left\{\left(-\frac{\mathrm{d}U}{\mathrm{d}x}\right)\Phi - \frac{1}{2v}\,U^2 + \tfrac{1}{2}w\Phi^2 - q\Phi\right\}\mathrm{d}x + \beta U(b) - \alpha U(a). \tag{1.4.8}$$

Let us now choose a trial function Φ which satisfies the boundary conditions (1.4.3), and determine $U = Y(\Phi)$ so that (1.4.4) holds identically. Then

$$Y(\Phi) = v\,\frac{\mathrm{d}\Phi}{\mathrm{d}x}. \tag{1.4.9}$$

Putting this in (1.4.7) we obtain the functional $J(\Phi)$ defined in (1.3.18)

$$J(\Phi) = I(v\Phi', \Phi) = \tfrac{1}{2}\int_a^b \{v(\Phi')^2 + w\Phi^2 - 2q\Phi\}\,\mathrm{d}x. \tag{1.4.10}$$

We note that this expression is identical to the functional $E(\Phi)$ in equation (1.2.19) of the example in section 1.2. Since H is quadratic, we see from theorem 1.3.2 that

$$J(\Phi) = I(u, \varphi) + \delta^2 J, \qquad (1.4.11)$$

where
$$\delta^2 J = \frac{1}{2} \int_a^b \left\{ (Y(\Phi) - u)^2 \frac{1}{v} + (\Phi - \varphi)^2 w \right\} dx. \qquad (1.4.12)$$

Since $v > 0$ and $w > 0$ here, we see that $\delta^2 J$ in (1.4.12) is non-negative, and hence we obtain the minimum principle

$$J(\Phi) \geqslant I(u, \varphi). \qquad (1.4.13)$$

This is the result obtained earlier in equation (1.2.22).

To investigate the complementary variational principle, we choose a trial function U, and determine $\Phi = \Theta(U)$ so that (1.4.5) holds identically. Then, since w is assumed to be non-zero,

$$\Theta(U) = \frac{1}{w} (q + U'). \qquad (1.4.14)$$

Setting this in (1.4.8) we obtain the functional $G(U)$ defined by (1.3.29):

$$G(U) = -\frac{1}{2} \int_a^b \left\{ \frac{1}{v} U^2 + \frac{1}{w} (q + U')^2 \right\} dx + \beta U(b) - \alpha U(a). \qquad (1.4.15)$$

Using theorem 1.3.3 we see that

$$G(U) = I(u, \varphi) + \delta^2 G, \qquad (1.4.16)$$

where
$$\delta^2 G = -\frac{1}{2} \int_a^b \left\{ (U - u)^2 \frac{1}{v} + (\Theta(U) - \varphi)^2 w \right\} dx. \qquad (1.4.17)$$

The formula in (1.4.17) is non-positive since $v > 0$ and $w > 0$, and hence (1.4.16) leads to the maximum principle

$$G(U) \leqslant I(u, \varphi). \qquad (1.4.18)$$

We may combine the results in (1.4.13) and (1.4.18) to obtain

$$G(U) \leqslant I(u, \varphi) \leqslant J(\Phi). \qquad (1.4.19)$$

Consequently, in this example the complementary variational principles (1.4.11) and (1.4.16) provide upper and lower bounds for the solution $I(u, \varphi)$ of the variational problem. We note that

$$I(u, \varphi) = -\tfrac{1}{2} \int_a^b q\varphi \, dx + \tfrac{1}{2} [v\varphi\varphi']_a^b, \qquad (1.4.20)$$

which follows from (1.4.7). Apart from the boundary term in (1.4.20), we see that upper and lower bounds have been obtained for a certain weighted average $\int q\varphi \, dx$ of the solution φ. In many applications the quantity $I(u, \varphi)$ is related to the energy of the physical system under investigation, and bounds for it may be of considerable interest.

Example 2.   In example 1 we assumed that w was nonzero. Let us now look at the case when $w = 0$. It is readily seen that, apart from equations (1.4.14) and (1.4.15), all the results derived in example 1 hold with $w = 0$. Consider then the modifications that are required to (1.4.14) and (1.4.15). These equations are concerned with the derivation of $G(U)$, which is obtained from $I(U, \Phi)$ by seeking $\Phi = \Theta(U)$ so that

$$-\frac{dU}{dx} = \frac{\partial H}{\partial \Phi} \tag{1.4.21}$$

holds identically. But (1.4.21) in this case reads

$$-\frac{dU}{dx} = q \quad \text{in} \quad a \leqslant x \leqslant b, \tag{1.4.22}$$

and this cannot be solved for $\Theta(U)$. Instead it represents a *constraint* on the trial function U. Setting such a trial U in (1.4.8) with $w = 0$ gives

$$G(U) = -\frac{1}{2}\int_a^b \frac{1}{v} \, U^2 \, dx + \beta U(b) - \alpha U(a). \tag{1.4.23}$$

Equations (1.4.22) and (1.4.23) are thus the modified forms of (1.4.14) and (1.4.15) when $w = 0$. We have included this example at this stage as it shows in a simple way how the canonical approach leads to constraints in an automatic fashion.

SUMMARY

This chapter has provided an elementary introduction to the basic ideas of complementary variational principles. The discussion was confined to the simplest problem in the calculus of variations, involving one-dimensional integrals with fixed end-points. The associated Euler–Lagrange variational theory was developed and extended to the canonical, or Hamiltonian, form. This canonical theory with its two independent functions and pair of symmetrical equations provided the setting for the derivation of complementary variational principles, which under certain circumstances lead to upper and lower bounds for the solution of the variational problem.

2

VARIATIONAL PRINCIPLES: SOME EXTENSIONS

2.1. A class of operators

So far the discussion has been limited to the simplest kind of variational problem for functions of one variable and fixed end-points. To make the theory more powerful we might seek to extend it in various ways. For example, we could introduce functions of several variables, or integrands L depending on operators other than $\mathrm{d}/\mathrm{d}x$. These extensions are of interest in mathematical physics, and we shall now turn our attention to them.

The background to this chapter concerns linear operators in real Hilbert space, and it is convenient at this point to recall some of the basic ideas [cf. 34].

Suppose that H_φ is a real Hilbert space. Then if φ_1 and φ_2 belong to H_φ, we can define an inner product $\langle \varphi_1, \varphi_2 \rangle$ on H_φ which is a real number that can be calculated when φ_1 and φ_2 are given. The inner product possesses the following properties:

(a) $\langle \varphi, \alpha\varphi_1 + \beta\varphi_2 \rangle = \alpha\langle \varphi, \varphi_1 \rangle + \beta\langle \varphi, \varphi_2 \rangle$,

where α and β are arbitrary real constants.

(b) $\langle \varphi_1, \varphi_2 \rangle = \langle \varphi_2, \varphi_1 \rangle$.
(c) $\langle \varphi, \varphi \rangle \geqslant 0$, with $\langle \varphi, \varphi \rangle = 0$ if and only if $\varphi = 0$.

One of the simplest definitions of inner product is

$$\langle \varphi, \psi \rangle = \int_a^b \varphi(x)\psi(x) \, \mathrm{d}x, \qquad (2.1.1)$$

corresponding to functions defined on the interval $a \leqslant x \leqslant b$. However, it is important to realize that there is some degree of flexibility in the definition of an inner product, and the choice is guided by the requirements of each individual problem.

An operator T is a transformation from one real Hilbert space H_φ to another space, say H_u. Thus we write

$$T\varphi = u, \quad \text{or} \quad T : H_\varphi \to H_u. \qquad (2.1.2)$$

We shall be concerned with linear operators, that is operators for which

$$T(\alpha\varphi+\beta\psi) = \alpha T\varphi+\beta T\psi, \tag{2.1.3}$$

where α and β are arbitrary real constants.

An important property of the operators that concern us is that, corresponding to a given linear operator $T:H_\varphi \to H_u$, there is a second operator $T^*:H_u \to H_\varphi$ such that

$$(u, T\varphi) = \langle T^*u, \varphi\rangle+[S(u, \varphi)] \tag{2.1.4}$$

for all functions φ, u in the domains of T, T^*. Here $(\,,\,)$ denotes an inner product defined on H_u and $[S(u, \varphi)]$ denotes boundary terms. The operator T^* is called the *adjoint* of T and $S(u, \varphi)$ is the corresponding *conjunct* of u and φ [cf. 34]. We shall write D_T for the domain of an operator T.

Some special cases of (2.1.4) are of particular interest to us and we now look at these in turn.

Example 1.   The first case corresponds to the basic operator appearing in Chapter I, namely

$$T = \frac{\mathrm{d}}{\mathrm{d}x}, \tag{2.1.5}$$

and we shall suppose that D_T consists of differentiable functions $\varphi(x)$ defined for $a \leqslant x \leqslant b$. If we set

$$(u, T\varphi) = \int_a^b u(x) \frac{\mathrm{d}}{\mathrm{d}x} \varphi(x) \, \mathrm{d}x \tag{2.1.6}$$

and integrate by parts, we can satisfy (2.1.4) by taking

$$\langle T^*u, \varphi\rangle = \int_a^b \left(-\frac{\mathrm{d}u}{\mathrm{d}x}\right)\varphi(x) \, \mathrm{d}x,$$

and

$$[S(u, \varphi)] = [u(x)\varphi(x)]_a^b. \tag{2.1.7}$$

Thus the adjoint operator in this example is

$$T^* = -\frac{\mathrm{d}}{\mathrm{d}x}, \tag{2.1.8}$$

and the domain D_{T^*} is $C_1(a, b)$.

Example 2.   Our second example follows from the divergence theorem in three dimensions:

$$\int_V \mathbf{u} \cdot \operatorname{grad} \varphi \, \mathrm{d}V = \int_V (-\operatorname{div} \mathbf{u})\varphi \, \mathrm{d}V + \int_{\partial V} \mathbf{u} \cdot \mathbf{n}\varphi \, \mathrm{d}B. \qquad (2.1.9)$$

We can define inner products so that (2.1.9) has the form

$$(\mathbf{u}, \operatorname{grad} \varphi) = \langle -\operatorname{div} \mathbf{u}, \varphi \rangle + [\mathbf{u} \cdot \mathbf{n}\varphi]. \qquad (2.1.10)$$

Comparison with (2.1.4) then shows that we can take

$$T = \operatorname{grad}, \qquad T^* = -\operatorname{div}, \qquad (2.1.11)$$

where D_T is the space of scalar functions which have partial derivatives in three dimensions and D_{T*} is the space of three-dimensional vector functions which possess partial derivatives.

Example 3.   Our third example is based on the identity

$$\int_V \mathbf{u} \cdot \operatorname{curl} \boldsymbol{\varphi} \, \mathrm{d}V = \int_V (\operatorname{curl} \mathbf{u}) \cdot \boldsymbol{\varphi} \, \mathrm{d}V + \int_{\partial V} \mathbf{u} \cdot (\mathbf{n} \wedge \boldsymbol{\varphi}) \, \mathrm{d}B. \qquad (2.1.12)$$

We define inner products so that (2.1.12) has the form

$$(\mathbf{u}, \operatorname{curl} \boldsymbol{\varphi}) = \langle \operatorname{curl} \mathbf{u}, \boldsymbol{\varphi} \rangle + [\mathbf{u} \cdot (\mathbf{n} \wedge \boldsymbol{\varphi})]. \qquad (2.1.13)$$

Comparison with (2.1.4) then shows that we can take

$$T = \operatorname{curl}, \qquad T^* = \operatorname{curl}, \qquad (2.1.14)$$

and the domains D_T and D_{T*} are both equal to the space of three-dimensional vector functions which possess partial derivatives.

Example 4.   Next we let T be an integral operator in the space L_2 of square-integrable functions. Thus we write

$$T\varphi(x) = \int_a^b k(x, y)\varphi(y) \, \mathrm{d}y, \qquad (2.1.15)$$

where the kernel $k(x, y)$ is any continuous function of x and y. If we define

$$(u, T\varphi) = \int_a^b u(x)T\varphi(x) \, \mathrm{d}x \qquad (2.1.16)$$

and use (2.1.15), we can satisfy (2.1.4) by taking $[S(u, \varphi)] = 0$ and

$$\langle T^*u, \varphi \rangle = \int_a^b \{T^*u(x)\}\varphi(x) \, \mathrm{d}x, \qquad (2.1.17)$$

where
$$T^*u(x) = \int_a^b k^*(x, y)u(y)\,\mathrm{d}y, \tag{2.1.18}$$

with
$$k^*(x, y) = k(y, x). \tag{2.1.19}$$

Thus we see that the kernel of the adjoint integral operator is the transpose of the kernel in (2.1.15).

Example 5.   Another interesting case is obtained by letting the operator T be an $m \times n$ matrix. To correspond with this choice we let $\varphi = [\varphi_i]$ and $u = [u_j]$ denote column vectors with n and m components respectively. We then define the inner products

$$(u, T\varphi) = u^t T\varphi \tag{2.1.20}$$

and
$$\langle T^*u, \varphi \rangle = \varphi^t T^t u \tag{2.1.21}$$

in terms of the usual scalar product of matrix multiplication, where t denotes the transpose. Thus the adjoint of T in this case is T^t, the transpose of T. There are no boundary terms in this example, so $S(u, \varphi) = 0$.

A class of operators

To end this section we note that the operators discussed in these examples possess a common property which may be expressed by the relation
$$(u, T\varphi) = \langle T^*u, \varphi \rangle + [u, \sigma_T \varphi], \tag{2.1.22}$$

where $\sigma_T : H_\varphi \to H_u$ is a certain operator. With the appropriate interpretation of the symbols in (2.1.22), the five examples correspond to

(i) $T = \dfrac{\mathrm{d}}{\mathrm{d}x} + v(x), \quad T^* = -\dfrac{\mathrm{d}}{\mathrm{d}x} + v(x), \ \sigma_T = 1,$

$\displaystyle (u, T\varphi) = \int_a^b u\left(\frac{\mathrm{d}}{\mathrm{d}x} + v\right)\varphi\,\mathrm{d}x, \quad \langle T^*u, \varphi \rangle = \int_a^b \left\{\left(-\frac{\mathrm{d}}{\mathrm{d}x} + v\right)u\right\}\varphi\,\mathrm{d}x,$

$[u, \sigma_T\varphi] = [u\varphi]_a^b.$

(ii) $T = \mathrm{grad}, \quad T^* = -\mathrm{div}, \quad \sigma_T = \mathbf{n}, \quad (u, T\varphi) = \displaystyle\int_V \mathbf{u}\,.\,\mathrm{grad}\ \varphi\,\mathrm{d}V,$

$\langle T^*u, \varphi \rangle = \displaystyle\int_V (-\mathrm{div}\ \mathbf{u})\varphi\,\mathrm{d}V, \quad [u, \sigma_T\varphi] = \displaystyle\int_{\partial V} \mathbf{u}\,.\,\mathbf{n}\varphi\,\mathrm{d}B.$

(iii) $T = \mathrm{curl}, \quad T^* = \mathrm{curl}, \quad \sigma_T = \mathbf{n}\wedge, \quad (u, T\varphi) = \displaystyle\int_V \mathbf{u}\,.\,\mathrm{curl}\ \boldsymbol{\varphi}\,\mathrm{d}V,$

$\langle T^*u, \varphi \rangle = \displaystyle\int_V (\mathrm{curl}\ \mathbf{u})\,.\,\boldsymbol{\varphi}\,\mathrm{d}V, \quad [u, \sigma_T\varphi] = \displaystyle\int_{\partial V} \mathbf{u}\,.\,\mathbf{n}\wedge\boldsymbol{\varphi}\,\mathrm{d}B.$

(iv) $T = \int k(x, y) \, dy, \quad T^* = \int k(y, x) \, dy, \quad \sigma_T = 0.$

$(u, T\varphi) = \int u(x)\{\int k(x, y)\varphi(y) \, dy\} \, dx,$

$\langle T^*u, \varphi \rangle = \int \{\int k(y, x)u(y) \, dy\}\varphi(x) \, dx.$

(v) $T = \text{matrix } M, T^* = \text{transpose of } M, \quad \sigma_T = 0,$

$(u, T\varphi) = u^t M\varphi, \quad \langle T^*u, \varphi \rangle = \varphi^t M^t u.$

We now turn to the variational theory associated with this class of operators.

2.2. Euler–Lagrange theory

Let T belong to the class of operators described in the previous section. Then the results of section 1.2 can be extended by considering a differentiable functional of the form

$$E(\Phi) = \int_V L(x, \Phi, T\Phi) \, dV, \qquad (2.2.1)$$

with given boundary condition

$$\sigma_T(\Phi - \varphi_B) = 0 \qquad \text{on} \quad \partial V. \qquad (2.2.2)$$

Here Φ belongs to the domain D_T of the operator T, V is some sufficiently simple region with boundary ∂V, x is a vector in V and φ_B specifies the behaviour of the function on the boundary.

Suppose now that the functional $E(\Phi)$ in (2.2.1) has an extremum at $\Phi = \varphi$. Then, as in section 1.2, we consider variations round φ

$$\Phi = \varphi + \epsilon \xi. \qquad (2.2.3)$$

The corresponding variation in $T\Phi$ is

$$T\Phi = T\varphi + \epsilon T\xi. \qquad (2.2.4)$$

Since $E(\Phi)$ is differentiable we can write

$$E(\varphi + \epsilon \xi) = E(\varphi) + \delta E(\varphi, \epsilon \xi) + \delta^2 E(\varphi, \epsilon \xi) + \ldots, \qquad (2.2.5)$$

where the first variation is

$$\delta E = \epsilon \int_V \left\{ \xi\left(\frac{\partial L}{\partial \Phi}\right)_\varphi + (T\xi)\left(\frac{\partial L}{\partial(T\Phi)}\right)_\varphi \right\} dV, \qquad (2.2.6)$$

and the second variation is

$$\delta^2 E = \tfrac{1}{2}\epsilon^2 \int_V \left\{ \xi \left(\frac{\partial^2 L}{\partial \Phi^2}\right)_\varphi \xi + 2\xi \left(\frac{\partial^2 L}{\partial \Phi\,\partial(T\Phi)}\right)_\varphi (T\xi) + (T\xi) \times \right.$$

$$\left. \times \left(\frac{\partial^2 L}{\partial(T\Phi)^2}\right)_\varphi (T\xi) \right\} \mathrm{d}V. \quad (2.2.7)$$

Now the operator T has an adjoint defined by (2.1.22), and using this definition we obtain an alternative expression for the first variation (2.2.6)

$$\delta E = \epsilon \int_V \xi \left[\frac{\partial L}{\partial \Phi} + T^* \left\{\frac{\partial L}{\partial(T\Phi)}\right\}\right]_\varphi \mathrm{d}V + \epsilon \int_{\partial V} (\sigma_T \xi)\left(\frac{\partial L}{\partial(T\Phi)}\right)_\varphi \mathrm{d}B. \quad (2.2.8)$$

Since for $\sigma_T \neq 0$ we are considering the class of functions Φ which satisfy the boundary condition (2.2.2), it follows that ξ vanishes on ∂V and hence (2.2.8) reduces to

$$\delta E = \epsilon \int_V \xi \left[\frac{\partial L}{\partial \Phi} + T^*\left\{\frac{\partial L}{\partial(T\Phi)}\right\}\right]_\varphi \mathrm{d}V. \quad (2.2.9)$$

For the functional $E(\Phi)$ in (2.2.1) to have an extremum, that is, be stationary at $\Phi = \varphi$, it is necessary that the first variation vanish. From (2.2.9) this means that

$$\int_V \xi \left[\frac{\partial L}{\partial \Phi} + T^*\left\{\frac{\partial L}{\partial(T\Phi)}\right\}\right]_\varphi \mathrm{d}V = 0. \quad (2.2.10)$$

Since $\xi \in D_T$ is arbitrary in the region V, it follows from (2.2.10) that

$$\left[\frac{\partial L}{\partial \Phi} + T^*\left\{\frac{\partial L}{\partial(T\Phi)}\right\}\right]_\varphi = 0 \quad \text{in} \quad V, \quad (2.2.11)$$

which is a generalized form of the Euler–Lagrange equation (1.2.11). We therefore can state

THEOREM 2.2.1. *The functional* $E(\Phi)$ *in* (2.2.1) *is stationary at* $\Phi = \varphi$ *where* φ *is the solution of*

$$\frac{\partial L}{\partial \Phi} + T^*\left\{\frac{\partial L}{\partial(T\Phi)}\right\} = 0 \quad \text{in} \quad V, \quad (2.2.12)$$

with
$$\sigma_T(\Phi - \varphi_B) = 0 \quad \text{on} \quad \partial V. \quad (2.2.13)$$

This is a generalized form [13, 62] *of the Euler–Lagrange variational principle.*

As noted in section 1.2, the procedure leading to equation (2.2.10) provides a necessary condition for an extremum, but, in general, one which is not sufficient. However, in many physical problems the existence of an extremum is often clear, and the generalized Euler–Lagrange equation (2.2.12) will be enough by itself to give a complete solution of the problem.

Assuming that we have found the function φ which makes $E(\Phi)$ stationary, we now wish to consider the nature of the extremum, that is, whether it is a maximum, a minimum, or a saddle-point. To do this we must look at the second variation defined in (2.2.7). If terms of $O(\epsilon^3)$ can be neglected, or if they vanish as is the case for quadratic L, it follows from (2.2.5) that

THEOREM 2.2.2. *A necessary condition for the functional $E(\Phi)$ to have a minimum at $\Phi = \varphi$ is that*

$$\delta^2 E(\varphi, \epsilon\xi) > 0 \qquad (2.2.14)$$

for all admissible $\xi \in D_T$, where $\delta^2 E$ is given by (2.2.7). Similarly, for a maximum at $\Phi = \varphi$, the condition is

$$\delta^2 E(\varphi, \epsilon\xi) < 0. \qquad (2.2.15)$$

When either of these results holds, we can obtain upper or lower bounds for the stationary value $E(\varphi)$ of the functional.

Example. To illustrate these results we consider a quadratic L, which is a generalized form of the one in equation (1.2.16). Thus, we take
$$L = \tfrac{1}{2}v(T\Phi)^2 + \tfrac{1}{2}w\Phi^2 - q\Phi, \qquad (2.2.16)$$
where v, w, and q may be functions of the vector x, with $v > 0$, $w \geqslant 0$ throughout the region V. The variational problem associated with (2.2.1) is then equivalent to the boundary value problem

$$T^*\{vT\varphi\} + w\varphi = q \qquad \text{in} \quad V \qquad (2.2.17)$$

$$\sigma_T(\varphi - \varphi_B) = 0 \qquad \text{on} \quad \partial V. \qquad (2.2.18)$$

These equations follow directly by setting (2.2.16) in theorem 2.2.1. Equation (2.2.17) is linear and, depending on the choice of T, it may be a differential, integral, or matrix equation.

From (2.2.5), (2.2.7), and (2.2.16) we find that the basic functional

$$E(\Phi) = \tfrac{1}{2}\int_V \{v(T\Phi)^2 + w\Phi^2 - 2q\Phi\} \, \mathrm{d}V \qquad (2.2.19)$$

may be expanded as

$$E(\Phi) = E(\varphi) + \tfrac{1}{2}\epsilon^2 \int_V \{w\xi^2 + v(T\xi)^2\}\, \mathrm{d}V \qquad (2.2.20)$$

on setting $\Phi = \varphi + \epsilon\xi$. Now $v > 0$ and $w \geqslant 0$ in V, and so from (2.2.20) we obtain the minimum principle

$$E(\Phi) \geqslant E(\varphi). \qquad (2.2.21)$$

If we take $T = \mathrm{grad}$, $v = 1$, $w = q = 0$, this result leads to the Dirichlet principle [cf. 59].

While this extended form of the Euler–Lagrange theory provides variational principles for a wide class of problems, it suffers from the limitation noted in section 1.2, namely that it leads in a natural way only to one-sided bounds. However, if a canonical approach to this extended theory is used, as was done in section 1.3 for the simple theory, it is found that in many cases both upper and lower bounds can be obtained for the stationary value. We now turn to this method, the potential use of which was first realized by Noble [55].

2.3. Canonical theory

As in the previous section we let T belong to the class of operators defined in section 2.1. Then the results of section 1.3 can be extended by taking instead of (1.3.6) the functional

$$I(U,\ \Phi) = \int_V \{UT\Phi - H(x,\ U,\ \Phi)\}\, \mathrm{d}V - \int_{\partial V} \{U\sigma_T\Phi - \Gamma(U,\ \Phi)\}\, \mathrm{d}B, \qquad (2.3.1)$$

where once again Γ is a boundary function determined by the particular form of boundary conditions that are required, Φ belongs to the domain D_T of the operator T, V is some sufficiently simple region with boundary ∂V, and x is a vector in V. It is assumed that the functional (2.3.1) is differentiable.

The argument now runs in parallel with that of section 1.3. Thus, we suppose that $I(U,\ \Phi)$ has an extremum at $U = u$, $\Phi = \varphi$. Then we consider variations round u and φ by setting

$$U = u + \epsilon\eta, \qquad \Phi = \varphi + \epsilon\xi \qquad (2.3.2)$$

in (2.3.1). This gives

$$I(U,\ \Phi) = I(u,\ \varphi) + \delta I + \delta^2 I + O(\epsilon^3), \qquad (2.3.3)$$

where
$$\delta I = \epsilon \int\limits_{V} \left\{ uT\xi + \eta T\varphi - \xi \left(\frac{\partial H}{\partial \Phi}\right)_{u,\varphi} - \eta \left(\frac{\partial H}{\partial U}\right)_{u,\varphi} \right\} \mathrm{d}V +$$

$$+ \epsilon \int\limits_{\partial V} \left\{ \eta \left(\frac{\partial \Gamma}{\partial U}\right)_{u,\varphi} + \xi \left(\frac{\partial \Gamma}{\partial \Phi}\right)_{u,\varphi} - u\sigma_T \xi - \eta \sigma_T \varphi \right\} \mathrm{d}B, \quad (2.3.4)$$

$$= \epsilon \int\limits_{V} \left\{ (T^*u)\xi + \eta T\varphi - \xi \left(\frac{\partial H}{\partial \Phi}\right)_{u,\varphi} - \eta \left(\frac{\partial H}{\partial U}\right)_{u,\varphi} \right\} \mathrm{d}V +$$

$$+ \epsilon \int\limits_{\partial V} \left\{ \eta \left(\frac{\partial \Gamma}{\partial U}\right)_{u,\varphi} + \xi \left(\frac{\partial \Gamma}{\partial \Phi}\right)_{u,\varphi} - \eta \sigma_T \varphi \right\} \mathrm{d}B, \quad (2.3.5)$$

by (2.1.22), and where

$$\delta^2 I = \tfrac{1}{2}\epsilon^2 \int\limits_{V} \left\{ 2\eta T\xi - \eta \left(\frac{\partial^2 H}{\partial U^2}\right)_{u,\varphi} \eta - 2\eta \left(\frac{\partial^2 H}{\partial U\,\partial \Phi}\right)_{u,\varphi} \xi - \xi \left(\frac{\partial^2 H}{\partial \Phi^2}\right)_{u,\varphi} \xi \right\} \mathrm{d}V -$$

$$- \tfrac{1}{2}\epsilon^2 \int\limits_{\partial V} \left\{ 2\eta \sigma_T \xi - \eta \left(\frac{\partial^2 \Gamma}{\partial U^2}\right)_{u,\varphi} \eta - 2\eta \left(\frac{\partial^2 \Gamma}{\partial U\,\partial \Phi}\right)_{u,\varphi} \xi - \xi \left(\frac{\partial^2 \Gamma}{\partial \Phi^2}\right)_{u,\varphi} \xi \right\} \mathrm{d}B,$$

$$(2.3.6)$$

or

$$\delta^2 I = \tfrac{1}{2}\epsilon^2 \int\limits_{V} \left\{ 2(T^*\eta)\xi - \eta \left(\frac{\partial^2 H}{\partial U^2}\right)_{u,\varphi} \eta - 2\eta \left(\frac{\partial^2 H}{\partial U\,\partial \Phi}\right)_{u,\varphi} \xi - \xi \left(\frac{\partial^2 H}{\partial \Phi^2}\right)_{u,\varphi} \xi \right\} \mathrm{d}V +$$

$$+ \tfrac{1}{2}\epsilon^2 \int\limits_{\partial V} \left\{ \eta \left(\frac{\partial^2 \Gamma}{\partial U^2}\right)_{u,\varphi} \eta + 2\eta \left(\frac{\partial^2 \Gamma}{\partial U\,\partial \Phi}\right)_{u,\varphi} \xi + \xi \left(\frac{\partial^2 \Gamma}{\partial \Phi^2}\right)_{u,\varphi} \xi \right\} \mathrm{d}B. \quad (2.3.7)$$

For the functional (2.3.1) to be stationary at $U = u$, $\Phi = \varphi$, it is necessary that $\delta I = 0$. From (2.3.5) this means that

$$\int\limits_{V} \left\{ \eta \left(T\Phi - \frac{\partial H}{\partial U}\right)_{u,\varphi} + \xi \left(T^*U - \frac{\partial H}{\partial \Phi}\right)_{u,\varphi} \right\} \mathrm{d}V +$$

$$+ \int\limits_{\partial V} \left\{ \eta \left(\frac{\partial \Gamma}{\partial U} - \sigma_T \Phi\right)_{u,\varphi} + \xi \left(\frac{\partial \Gamma}{\partial \Phi}\right)_{u,\varphi} \right\} \mathrm{d}B = 0. \quad (2.3.8)$$

We now choose the boundary function Γ to be

$$\Gamma(U, \Phi) = U\sigma_T \varphi_B. \quad (2.3.9)$$

With this choice of Γ the basic functional (2.3.1) becomes

$$I(U, \Phi) = \int\limits_{V} \{ UT\Phi - H(x, U, \Phi) \} \, \mathrm{d}V - \int\limits_{\partial V} U\sigma_T (\Phi - \varphi_B) \, \mathrm{d}B, \quad (2.3.10)$$

$$= \int\limits_{V} \{ (T^*U)\Phi - H(x, U, \Phi) \} \, \mathrm{d}V + \int\limits_{\partial V} U\sigma_T \varphi_B \, \mathrm{d}B, \quad (2.3.11)$$

and condition (2.3.8) reads

$$\int_V \left\{ \eta \left(T\Phi - \frac{\partial H}{\partial U} \right)_{u,\varphi} + \xi \left(T^*U - \frac{\partial H}{\partial \Phi} \right)_{u,\varphi} \right\} dV +$$

$$+ \int_{\partial V} \eta \sigma_T (\varphi_B - \Phi)_{u,\varphi} \, dB = 0. \quad (2.3.12)$$

From (2.3.12) we therefore obtain the following principle for the functional $I(U, \Phi)$:

THEOREM 2.3.1. *The functional $I(U, \Phi)$ in (2.3.10) and (2.3.11) is stationary at u, φ, where u, φ are the solutions of the boundary-value problem*

$$T\Phi = \frac{\partial H}{\partial U} \quad in \quad V, \quad (2.3.13)$$

$$T^*U = \frac{\partial H}{\partial \Phi} \quad in \quad V, \quad (2.3.14)$$

$$\sigma_T(\Phi - \varphi_B) = 0 \quad on \quad \partial V. \quad (2.3.15)$$

Equations (2.3.13) and (2.3.14) provide a generalized form of the canonical Euler equations. As noted earlier in Chapter I, other boundary conditions can also be included in the theory [cf. 12, 14] by choosing appropriate forms for Γ, but we shall concentrate here on the boundary condition (2.3.15).

Now we turn to the derivation of maximum and minimum principles in which we say that under certain circumstances $I(u, \varphi)$ is greater than or less than $I(U, \varphi)$.

Suppose first that we choose a trial function Φ and assume that

$$T\Phi = \frac{\partial H}{\partial U} \quad \text{in } V \text{ and on } \partial V \quad (2.3.16)$$

has the solution $U = Y(\Phi)$. This U is then substituted in (2.3.10) and we write the resulting functional as

$$I(Y(\Phi), \Phi) = J(\Phi). \quad (2.3.17)$$

From theorem 2.3.1 it is clear that the functional $J(\Phi)$ determined in this way is stationary at φ, and we expand

$$J(\Phi) = I(u, \varphi) + \delta^2 J + O(\epsilon^3), \quad (2.3.18)$$

3

where from (2.3.6) and (2.3.9)

$$\delta^2 J = \tfrac{1}{2}\epsilon^2 \int_V \left\{ 2\eta T\xi - \eta \left(\frac{\partial^2 H}{\partial U^2}\right)_{u,\varphi} \eta - 2\eta \left(\frac{\partial^2 H}{\partial U\,\partial\Phi}\right)_{u,\varphi} \xi - \xi \left(\frac{\partial^2 H}{\partial\Phi^2}\right)_{u,\varphi} \xi \right\} dV -$$

$$- \epsilon^2 \int_{\partial V} \eta\sigma_T\xi \, dB, \quad (2.3.19)$$

in which $\epsilon\xi = \Phi - \varphi$ and $\epsilon\eta = Y(\Phi) - u$. From (2.3.16) we find that

$$T\xi = \left(\frac{\partial^2 H}{\partial U^2}\right)_{u,\varphi} \eta + \left(\frac{\partial^2 H}{\partial U\,\partial\Phi}\right)_{u,\varphi} \xi + O(\epsilon), \quad (2.3.20)$$

and so (2.3.19) may be simplified to

$$\delta^2 J = \tfrac{1}{2}\int_V \left\{ (Y(\Phi)-u)\left(\frac{\partial^2 H}{\partial U^2}\right)_{u,\varphi} (Y(\Phi)-u) - (\Phi-\varphi)\times \right.$$

$$\left. \times \left(\frac{\partial^2 H}{\partial\Phi^2}\right)_{u,\varphi} (\Phi-\varphi) \right\} dV - \int_{\partial V} (Y(\Phi)-u)\sigma_T(\Phi-\varphi_B) \, dB. \quad (2.3.21)$$

At this point we depart from the procedure in section 1.3 and assume nothing about Φ on the boundary ∂V. In most cases it is possible to satisfy $\sigma_T(\Phi-\varphi_B) = 0$ on ∂V, but in some cases this is not so and therefore we leave the choice open at present. These results concerning the functional $J(\Phi)$ are put together in

THEOREM 2.3.2. *The functional $J(\Phi)$ defined by (2.3.17) is stationary as Φ varies round φ. For Φ sufficiently close to φ,*

$$J(\Phi) \leqslant I(u, \varphi) \quad if \quad \delta^2 J \leqslant 0 \quad (2.3.22)$$

or $$J(\Phi) \geqslant I(u, \varphi) \quad if \quad \delta^2 J \geqslant 0 \quad (2.3.23)$$

where $\delta^2 J$ is given by (2.3.21).

The complementary variational principle is obtained by considering the alternative form of $I(U, \Phi)$ given by (2.3.11),

$$I(U, \Phi) = \int_V \{(T^*U)\Phi - H(x, U, \Phi)\} \, dV + \int_{\partial V} U\sigma_T\varphi_B \, dB. \quad (2.3.11)$$

In this case we first guess U and then assume that

$$T^*U = \frac{\partial H}{\partial\Phi} \quad in \quad V \quad (2.3.24)$$

has the solution $\Phi = \Theta(U)$. This Φ is then substituted in (2.3.11) and we write the resulting functional as

$$I(U, \Theta(U)) = G(U). \tag{2.3.25}$$

It is clear from theorem 2.3.1 that the functional $G(U)$ determined in this way is stationary at u, and we write

$$G(U) = I(u, \varphi) + \delta^2 G + O(\epsilon^3), \tag{2.3.26}$$

where, from (2.3.7) and (2.3.9),

$$\delta^2 G = \tfrac{1}{2}\epsilon^2 \int_V \left\{ 2(T^*\eta)\xi - \eta\left(\frac{\partial^2 H}{\partial U^2}\right)_{u,\varphi}\eta - 2\eta\left(\frac{\partial^2 H}{\partial U\,\partial \Phi}\right)_{u,\varphi}\xi - \right.$$
$$\left. -\xi\left(\frac{\partial^2 H}{\partial \Phi^2}\right)_{u,\varphi}\xi\right\}\,\mathrm{d}V, \quad (2.3.27)$$

in which $\epsilon\xi = \Theta(U) - \varphi$ and $\epsilon\eta = U - u$. From (2.3.24) we find that

$$T^*\eta = \left(\frac{\partial^2 H}{\partial \Phi\,\partial U}\right)_{u,\varphi}\eta + \left(\frac{\partial^2 H}{\partial \Phi^2}\right)_{u,\varphi}\xi + O(\epsilon), \tag{2.3.28}$$

and so (2.3.27) becomes

$$\delta^2 G = -\frac{1}{2}\int_V\left\{(U-u)\left(\frac{\partial^2 H}{\partial U^2}\right)_{u,\varphi}(U-u) - (\Theta(U)-\varphi)\left(\frac{\partial^2 H}{\partial \Phi^2}\right)_{u,\varphi}\times\right.$$
$$\left.\times(\Theta(U)-\varphi)\right\}\,\mathrm{d}V. \quad (2.3.29)$$

These results for G are collected together in

THEOREM 2.3.3. *The functional $G(U)$ defined by (2.3.25) is stationary as U varies round u. For U sufficiently close to u,*

$$G(U) \leqslant I(u, \varphi) \quad if \quad \delta^2 G \leqslant 0, \tag{2.3.30}$$
$$or \qquad G(U) \geqslant I(u, \varphi) \quad if \quad \delta^2 G \geqslant 0 \tag{2.3.31}$$

where $\delta^2 G$ is given by (2.3.29). This principle is complementary to the one given in theorem 2.3.2.

2.4. Complementary variational principles

Having obtained the results in theorems 2.3.2 and 2.3.3, we are now able to state the main result of this monograph, the theorem of Noble [55] on complementary variational principles.

THEOREM 2.4.1. *With the definitions and assumptions of theorems* 2.3.2 *and* 2.3.3,

$$G(U) \leqslant I(u, \varphi) \leqslant J(\Phi) \qquad if \quad \delta^2 G \leqslant 0, \, \delta^2 J \geqslant 0, \qquad (2.4.1)$$

equality holding when U *and* Φ *are the exact solutions* u, φ *of the variational problem* (2.3.10). *This result also holds if all the inequality signs are reversed.*

Formula (2.4.1) is the reason why the variational principles (2.3.18) and (2.3.26) are said to be complementary.

The maximum and minimum principles in theorems 2.3.2 and 2.3.3 give complementary upper and lower bounds for $I(u, \varphi)$, provided that $\delta^2 G$ and $\delta^2 J$ do not have the same sign, and provided that U and Φ are sufficiently close to u and φ. The pair of functions (u, φ) furnishes the exact solution of the problem described by equations (2.3.13)–(2.3.15).

2.5. Abstract formalism

Before we apply these results on complementary variational principles to certain boundary-value problems, we shall introduce a slightly more abstract form of the theory [cf. 66, 45]. The advantage of this form is that it allows us to deal with variational problems involving differential, integral, and matrix operators in a unified and succinct manner.

We consider first the Euler–Lagrange theory on a real Hilbert space H_φ with inner product $<, >$. Let $E(\varphi)$ be a functional on H_φ which is twice differentiable in the sense of Fréchet [82]. Then we can expand

$$E(\varphi + \epsilon \xi) = E(\varphi) + \delta E + \delta^2 E + O(\epsilon^3), \qquad (2.5.1)$$

where the first variation

$$\delta E = \left\langle \epsilon \xi, \frac{\delta E}{\delta \varphi} \right\rangle, \qquad (2.5.2)$$

and the second variation

$$\delta^2 E = \frac{1}{2} \left\langle \epsilon \xi, \frac{\delta^2 E}{\delta \varphi^2} \, \epsilon \xi \right\rangle. \qquad (2.5.3)$$

Here $\delta E / \delta \varphi \in H_\varphi$ is the functional derivative of E, or gradient vector [cf. 28], and $\delta^2 E / \delta \varphi^2 : H_\varphi \to H_\varphi$ is an operator defined by the expansion (2.5.1) and (2.5.3).

Example. To illustrate the notion of functional derivative we consider the quadratic functional $E(\varphi)$ on H_φ given by

$$E(\varphi) = \tfrac{1}{2}\langle \varphi, A\varphi \rangle, \tag{2.5.4}$$

where $A : H_\varphi \to H_\varphi$ is a symmetric linear operator. Then

$$E(\varphi + \epsilon\xi) = \tfrac{1}{2}\langle \varphi + \epsilon\xi, A(\varphi + \epsilon\xi)\rangle$$

$$= \tfrac{1}{2}\langle \varphi, A\varphi\rangle + \langle \epsilon\xi, A\varphi\rangle + \tfrac{1}{2}\langle \epsilon\xi, A\epsilon\xi\rangle. \tag{2.5.5}$$

But from (2.5.1) we have

$$E(\varphi + \epsilon\xi) = E(\varphi) + \left\langle \epsilon\xi, \frac{\delta E}{\delta\varphi}\right\rangle + \frac{1}{2}\left\langle \epsilon\xi, \frac{\delta^2 E}{\delta\varphi^2}\,\epsilon\xi\right\rangle + O(\epsilon^3), \tag{2.5.6}$$

and hence we see that for this example the first functional derivative is

$$\frac{\delta E}{\delta\varphi} = A\varphi, \tag{2.5.7}$$

which is a vector in H_φ, and the second functional derivative is

$$\frac{\delta^2 E}{\delta\varphi^2} = A, \tag{2.5.8}$$

which is an operator from H_φ to H_φ, and all higher functional derivatives vanish.

We now return to the functional $E(\varphi)$ in (2.5.1). Suppose that $E(\Phi)$ has an extremum at $\Phi = \varphi$. Then a necessary condition for $E(\Phi)$ to be stationary at $\Phi = \varphi$ is that the first variation vanish. Using (2.5.2) we can therefore state

THEOREM 2.5.1. *The functional $E(\Phi)$ is stationary at $\Phi = \varphi$ where φ is the solution of*

$$\frac{\delta E}{\delta\varphi}(\varphi) = 0. \tag{2.5.9}$$

This is an abstract form of the Euler–Lagrange equation.

In addition, we can see from (2.5.1) that if $\delta^2 E/\delta\varphi^2$ is a positive operator, then

$$E(\varphi) \leqslant E(\Phi) \tag{2.5.10}$$

for suitable functions $\Phi \in H_\varphi$ sufficiently close to φ. So in this case a minimum principle is obtained which gives a one-sided bound for the solution $E(\varphi)$ of the variational problem. If $\delta^2 E/\delta\varphi^2$ is negative, the result is a maximum principle. Thus, in these cases, an upper or a lower bound (but not both) for $E(\varphi)$ is obtained. This corresponds to the standard Euler–Lagrange approach. To derive both upper and lower bounds we turn to the canonical theory.

As in section 2.1 we let H_u and H_φ be real Hilbert spaces with inner products $(,)$ and $\langle,\rangle$ respectively. Also we let $T:H_\varphi \to H_u$ be a linear operator with adjoint $T^*:H_u \to H_\varphi$ defined by the relation (2.1.22)

$$(u,\ T\varphi) = \langle T^*u,\ \varphi\rangle + [u,\ \sigma_T\varphi] \tag{2.5.11}$$

$$= \langle T^*u,\ \varphi\rangle + [\sigma_T^*u,\ \varphi] \tag{2.5.12}$$

for all $u \in H_u$, $\varphi \in H_\varphi$, where $\sigma_T^*:H_u \to H_\varphi$ is the adjoint of the linear operator $\sigma_T:H_\varphi \to H_u$. Here we can, for example, identify the inner products as

$$(u,\ T\varphi) = \int_V uT\varphi \ \mathrm{d}V,$$

$$\langle T^*u,\ \varphi\rangle = \int_V (T^*u)\varphi \ \mathrm{d}V,$$

$$[u,\ \sigma_T\varphi] = \int_{\partial V} u\sigma_T\varphi \ \mathrm{d}B. \tag{2.5.13}$$

For an unbounded operator such as grad it is understood that the inner product is defined for a corresponding bounded linear operator T on a suitable domain H_φ. In terms of these inner products we can therefore write the basic functional $I(u,\ \varphi)$ of (2.3.10) and (2.3.11) as

$$I(u,\ \varphi) = (u,\ T\varphi) - W(u,\ \varphi) - [u,\ \sigma_T(\varphi - \varphi_B)], \tag{2.5.14}$$

$$= \langle T^*u,\ \varphi\rangle - W(u,\ \varphi) + [u,\ \sigma_T\varphi_B]. \tag{2.5.15}$$

Here the functional $W(u,\ \varphi)$ is related to the generalized Hamiltonian $H(x,\ u,\ \varphi)$ through

$$W(u,\ \varphi) = \int_V H(x,\ u,\ \varphi) \ \mathrm{d}V. \tag{2.5.16}$$

The functionals in (2.5.14) and (2.5.15) are defined on the product space $H = H_u \times H_\varphi$.

Suppose that $W(u,\ \varphi)$ defined on H is twice differentiable in the sense of Fréchet [cf. 82]. Then consider the expansion

$$I(u+\epsilon\eta,\ \varphi+\epsilon\xi) = I(u,\ \varphi) + \delta I + \delta^2 I + O(\epsilon^3), \tag{2.5.17}$$

where the first variation

$$\delta I = \left(\epsilon\eta,\ T\varphi - \frac{\delta W}{\delta u}\right) + \left\langle \epsilon\xi,\ T^*u - \frac{\delta W}{\delta\varphi}\right\rangle - [\epsilon\eta,\ \sigma_T(\varphi - \varphi_B)], \tag{2.5.18}$$

and the second variation

$$\delta^2 I = -\frac{1}{2}\left(\epsilon\eta, \frac{\delta^2 W}{\delta u^2}\,\epsilon\eta\right) + \frac{1}{2}\left(\epsilon\eta, \left(T - \frac{\delta^2 W}{\delta u\,\delta\varphi}\right)\epsilon\xi\right) +$$

$$+ \frac{1}{2}\left\langle \epsilon\xi, \left(T^* - \frac{\delta^2 W}{\delta\varphi\,\delta u}\right)\epsilon\eta\right\rangle - \frac{1}{2}\left\langle \epsilon\xi, \frac{\delta^2 W}{\delta\varphi^2}\,\epsilon\xi\right\rangle - \tfrac{1}{2}[\epsilon\eta,\ \sigma_T\epsilon\xi]. \quad (2.5.19)$$

These results follow from an extension of (2.5.1) to a product space. In (2.5.18) $\delta W/\delta u \in H_u$ and $\delta W/\delta\varphi \in H_\varphi$ are gradient vectors, and the operators in (2.5.19) are symmetric bilinear operators such that

$$\left(u_1, \frac{\delta^2 W}{\delta u^2}\,u_2\right) = \left(\frac{\delta^2 W}{\delta u^2}\,u_1, u_2\right),$$

$$\left(u_1, \frac{\delta^2 W}{\delta u\,\delta\varphi}\,\varphi_1\right) = \left\langle \frac{\delta^2 W}{\delta\varphi\,\delta u}\,u_1, \varphi_1\right\rangle,$$

$$\left\langle \varphi_1, \frac{\delta^2 W}{\delta\varphi^2}\,\varphi_2\right\rangle = \left\langle \frac{\delta^2 W}{\delta\varphi^2}\,\varphi_1, \varphi_2\right\rangle, \quad (2.5.20)$$

for all $u_1,\ u_2 \in H_u$ and $\varphi_1,\ \varphi_2 \in H_\varphi$. The convention here is that $\delta^2 W/\delta u\delta\varphi = (\delta/\delta\varphi)(\delta W/\delta u)$.

A necessary condition for the functional $I(u, \varphi)$ to be stationary at (u, φ) is that $\delta I = 0$ at (u, φ). From (2.5.18) we therefore obtain the following principle.

THEOREM 2.5.2. *The functional* $I(u, \varphi)$ *in* (2.5.14) *and* (2.5.15) *is stationary at* u, φ, *where* u, φ *are the solutions of the boundary value problem*

$$T\varphi = \frac{\delta W}{\delta u} \qquad in \quad V, \quad (2.5.21)$$

$$T^*u = \frac{\delta W}{\delta\varphi} \qquad in \quad V, \quad (2.5.22)$$

$$\sigma_T(\varphi - \varphi_B) = 0 \qquad on \quad \partial V. \quad (2.5.23)$$

The terms in (2.5.21) *and* (2.5.23) *are elements of* H_u, *and the terms in* (2.5.22) *are elements of* H_φ. *The equations* (2.5.21) *and* (2.5.22) *are an abstract form of the canonical Euler, or Hamilton, equations.*

Now we turn to the derivation of extremum principles. Such principles cannot be obtained, however, from equation (2.5.17) because the functions $u + \epsilon\eta$ and $\varphi + \epsilon\xi$ are arbitrary and independent. Some restriction on these functions is necessary and involves making one function dependent on the other in a certain way.

Assume that the first canonical equation

$$T\Phi = \frac{\delta W}{\delta U}(U, \Phi) \qquad (2.5.24)$$

has the solution $U = Y(\Phi)$. Then we define the functional

$$J(\Phi) = I(Y(\Phi), \Phi), \qquad (2.5.25)$$

which can be expanded about $\Phi = \varphi$ to give

$$J(\Phi) = I(u, \varphi) + \delta^2 J + O(\epsilon^3), \qquad (2.5.26)$$

where

$$\delta^2 J = \delta^2 I(Y(\Phi), \Phi). \qquad (2.5.27)$$

Thus $J(\Phi)$ is stationary at $\Phi = \varphi$.

Also, writing $\Phi = \varphi + \epsilon\xi$ in (2.5.24) and using $T\varphi = \delta W/\delta u$, we obtain

$$\left(T - \frac{\delta^2 W}{\delta u \, \delta\varphi}\right)\epsilon\xi = \frac{\delta^2 W}{\delta u^2}\,\epsilon\eta + O(\epsilon^2). \qquad (2.5.28)$$

Setting this in (2.5.19) we derive a simple expression for $\delta^2 J$, namely

$$\delta^2 J = \frac{1}{2}\left(\epsilon\eta, \frac{\delta^2 W}{\delta u^2}\,\epsilon\eta\right) - \frac{1}{2}\left\langle \epsilon\xi, \frac{\delta^2 W}{\delta\varphi^2}\,\epsilon\xi \right\rangle - [\epsilon\eta, \sigma_T\epsilon\xi]. \qquad (2.5.29)$$

Another principle can be derived by supposing that the second canonical equation

$$T^*U = \frac{\delta W}{\delta\Phi}(U, \Phi) \qquad (2.5.30)$$

has the solution $\Phi = \Theta(U)$. Then we define the functional

$$G(U) = I(U, \Theta(U)), \qquad (2.5.31)$$

which can be expanded about $U = u$ to give

$$G(U) = I(u, \varphi) + \delta^2 G + O(\epsilon^3), \qquad (2.5.32)$$

where

$$\delta^2 G = \delta^2 I(U, \Theta(U)). \qquad (2.5.33)$$

Thus $G(U)$ is stationary at $U = u$.

Also, writing $U = u + \epsilon\eta$ in (2.5.30) and using $T^*u = \delta W/\delta\varphi$, we obtain

$$\left(T^* - \frac{\delta^2 W}{\delta\varphi \, \delta u}\right)\epsilon\eta = \frac{\delta^2 W}{\delta\varphi^2}\,\epsilon\xi + O(\epsilon^2). \qquad (2.5.34)$$

Setting this in (2.5.19) we find that

$$\delta^2 G = -\frac{1}{2}\left(\epsilon\eta, \frac{\delta^2 W}{\delta u^2}\,\epsilon\eta\right) + \frac{1}{2}\left\langle \epsilon\xi, \frac{\delta^2 W}{\delta\varphi^2}\,\epsilon\xi \right\rangle. \qquad (2.5.35)$$

We can now state the theorem on complementary variational principles.

THEOREM 2.5.3. *Let (u, φ) be a solution of* (2.5.21)–(2.5.23), *and suppose that it is possible to obtain solutions of the equations* (2.5.24) *and* (2.5.30) *in the form*

$$U = Y(\Phi) \quad and \quad \Phi = \Theta(U). \qquad (2.5.36)$$

Then, for (U, Φ) *sufficiently close to* (u, φ) *the upper and lower bounds*

$$G(U) \equiv I(U, \Theta(U)) \leqslant I(u, \varphi) \leqslant I(Y(\Phi), \Phi) \equiv J(\Phi) \qquad (2.5.37)$$

hold, provided that

$$\delta^2 J = \frac{1}{2}\left(Y-u, \frac{\delta^2 W}{\delta u^2}\,(Y-u)\right) - \frac{1}{2}\left\langle \Phi-\varphi, \frac{\delta^2 W}{\delta\varphi^2}\,(\Phi-\varphi)\right\rangle -$$

$$-[Y-u, \sigma_T(\Phi-\varphi_B)] \geqslant 0, \qquad (2.5.38)$$

and $\qquad \delta^2 G = -\frac{1}{2}\left\{\left(U-u, \frac{\delta^2 W}{\delta u^2}\,(U-u)\right) - \left\langle \Theta-\varphi, \frac{\delta^2 W}{\delta\varphi^2}\,(\Theta-\varphi)\right\rangle\right\} \leqslant 0.$

$$(2.5.39)$$

If the inequality signs are reversed in (2.5.38) *and* (2.5.39), *then of course they are reversed in* (2.5.37).

A detailed proof of this result can be given on the lines indicated by Rall [66].

Before we leave the abstract formalism, we note for future reference that to find the functional $E(\varphi)$ such that $\delta E/\delta\varphi = B(\varphi)$, we use the result of direct integration [82], which gives

$$E(\varphi) = \int_0^1 \langle\varphi-\varphi_0, B\{\varphi_0+t(\varphi-\varphi_0)\}\rangle\,\mathrm{d}t + E(\varphi_0), \qquad (2.5.40)$$

where φ_0 is an arbitrary vector in H_φ. For example, if $B(\varphi) = A\varphi$ as in (2.5.7), then (2.5.40) with $\varphi_0 = 0$ gives $E(\varphi) = \frac{1}{2}\langle\varphi, A\varphi\rangle$ which is just the functional in (2.5.4).

This completes the general development of the theory. In the remaining sections of the book we shall derive complementary variational principles for certain classes of linear and nonlinear boundary-value problems and apply the results to various problems in mathematical physics.

SUMMARY

A natural extension of the theory of Chapter I to cover a certain class of operator equations and associated variational problems has been described in this chapter. The Euler–Lagrange theory and its canonical form were developed for various differential, integral, and matrix operators in a unified manner. As in Chapter I, the canonical theory led to complementary variational principles in certain cases. An abstract treatment involving functional derivatives and gradient vectors was included in the last section to provide a succinct derivation of the basic results.

3

LINEAR APPLICATIONS

3.1. A class of linear problems

THE general variational theory of the previous chapter can now be applied to an important class of linear boundary-value problems described by the equations

$$A\varphi = f \quad \text{in} \quad V, \tag{3.1.1}$$

with
$$A = T^*T + Q, \tag{3.1.2}$$

subject to
$$\sigma_T(\varphi - \varphi_B) = 0 \quad \text{on} \quad \partial V. \tag{3.1.3}$$

Here $T : H_\varphi \to H_u$ and its adjoint $T^* : H_u \to H_\varphi$ are linear operators of the kind discussed in Chapter 2 such that

$$(u, T\varphi) = \langle T^*u, \varphi \rangle + [u, \sigma_T\varphi], \tag{3.1.4}$$

and Q is a symmetric positive operator, which means that

$$\langle \varphi_1, Q\varphi_2 \rangle = \langle Q\varphi_1, \varphi_2 \rangle \quad \text{for all } \phi_1, \phi_2 \text{ in } D_Q, \tag{3.1.5}$$

and
$$\langle \varphi, Q\varphi \rangle \geqslant 0 \quad \text{for all } \varphi \text{ in } D_Q. \tag{3.1.6}$$

D_A denotes the domain of the linear operator A which is dense in the real Hilbert space H_φ, and f is a given function in H_φ, and φ_B is a prescribed function on the boundary ∂V of the region V. It is readily seen that

$$\langle \varphi_1, A\varphi_2 \rangle = \langle A\varphi_1, \varphi_2 \rangle + [T\varphi_1, \sigma_T\varphi_2] - [T\varphi_2, \sigma_T\varphi_1] \tag{3.1.7}$$

for all φ_1, φ_2 in D_A, and

$$\langle \varphi, A\varphi \rangle = (T\varphi, T\varphi) + \langle \varphi, Q\varphi \rangle - [T\varphi, \sigma_T\varphi]. \tag{3.1.8}$$

Relation (3.1.7) shows that in a formal sense A is symmetric, and (3.1.8) shows that, for D_A restricted to functions φ such that $[T\varphi, \sigma_T\varphi] = 0$, the operator A is positive.

The class of problems in (3.1.1)–(3.1.3) contains many boundary-value problems of mathematical physics, and it is therefore convenient for later applications to develop the corresponding variational theory at this point. We shall assume that the problem in (3.1.1)–(3.1.3) has a solution; in which case we note that when A is positive-definite, that is

$\langle \varphi, A\varphi \rangle > 0$ for every $\varphi \neq 0$, the solution is unique. Approximate methods for equations of the form (3.1.1) with $A = T^*T$ were first considered by Kato [42].

3.2. Associated variational theory

We wish to apply the general canonical theory of Chapter 2 to the system (3.1.1)–(3.1.3), and to do this we must write (3.1.1) as a pair of canonical equations. An appropriate form is given by

$$T\varphi = u = \delta W/\delta u, \tag{3.2.1}$$

$$T^*u = f - Q\varphi = \delta W/\delta \varphi, \tag{3.2.2}$$

and by direct integration as in (2.5.40), we see that a suitable W is

$$W(u, \varphi) = \tfrac{1}{2}(u, u) + \langle f, \varphi \rangle - \tfrac{1}{2}\langle \varphi, Q\varphi \rangle. \tag{3.2.3}$$

Having found $W(u, \varphi)$ we can now obtain the potential or basic 'action' functional $I(u, \varphi)$ of the variational theory from (2.5.14) and (2.5.15). This is given by

$$I(U, \Phi) = (U, T\Phi) - \tfrac{1}{2}(U, U) - \langle f, \Phi \rangle + \tfrac{1}{2}\langle \Phi, Q\Phi \rangle - [U, \sigma_T(\Phi - \varphi_B)], \tag{3.2.4}$$

$$= \langle T^*U, \Phi \rangle - \tfrac{1}{2}(U, U) - \langle f, \Phi \rangle + \tfrac{1}{2}\langle \Phi, Q\Phi \rangle + [U, \sigma_T\varphi_B]. \tag{3.2.5}$$

As would be expected for a linear boundary-value problem, the corresponding potential $I(u, \varphi)$ is a quadratic functional, which in turn means that third and higher order variations are all identically zero.

If (u, φ) denotes the exact solution of (3.2.1) and (3.2.2) subject to (3.1.3), it follows from (3.2.4) or (3.2.5) that

$$I(u, \varphi) = -\tfrac{1}{2}\langle f, \varphi \rangle + \tfrac{1}{2}[T\varphi, \sigma_T\varphi_B] \tag{3.2.6}$$

is the solution of the variational problem. Apart from the boundary term we see that this quantity is a certain weighted average of the exact solution φ.

From theorem 2.5.3 we know that under certain circumstances it is possible to obtain upper and lower bounds for the value of $I(u, \varphi)$. We therefore derive the corresponding functionals G and J and examine their bounding properties.

First we derive J which by (2.5.37) is defined by

$$J(\Phi) = I(Y, \Phi), \tag{3.2.7}$$

it being assumed that the canonical equation $T\Phi = \delta W/\delta U$ has the solution

$$U = Y(\Phi). \tag{3.2.8}$$

From (3.2.1) this means that

$$Y(\Phi) = T\Phi, \tag{3.2.9}$$

and so

$$J(\Phi) = I(T\Phi, \Phi). \tag{3.2.10}$$

By (3.2.4) we therefore find that

$$J(\Phi) = \tfrac{1}{2}(T\Phi, T\Phi) - \langle f, \Phi \rangle + \tfrac{1}{2}\langle \Phi, Q\Phi \rangle - [T\Phi, \sigma_T(\Phi - \varphi_B)], \tag{3.2.11}$$

$$= \tfrac{1}{2}\langle \Phi, A\Phi \rangle - \langle f, \Phi \rangle - [T\Phi, \sigma_T(\tfrac{1}{2}\Phi - \varphi_B)]. \tag{3.2.12}$$

Since $J(\Phi)$ is quadratic, an expansion about the exact function φ gives

$$J(\Phi) = I(u, \varphi) + \delta^2 J(\Phi), \tag{3.2.13}$$

where from (2.5.38)

$$\delta^2 J = \frac{1}{2}\left(Y - u, \frac{\delta^2 W}{\delta u^2}(Y - u)\right) - \frac{1}{2}\left\langle \Phi - \varphi, \frac{\delta^2 W}{\delta \varphi^2}(\Phi - \varphi)\right\rangle -$$

$$-[Y - u, \sigma_T(\Phi - \varphi_B)]. \tag{3.2.14}$$

Now by (3.2.1) and (3.2.2) we can find the second derivatives of W

$$\frac{\delta^2 W}{\delta u^2} = 1, \qquad \frac{\delta^2 W}{\delta \varphi^2} = -Q. \tag{3.2.15}$$

Setting these in (3.2.14) and using $Y = T\Phi$, $u = T\varphi$, we can then write $\delta^2 J$ as

$$\delta^2 J = \tfrac{1}{2}(T\Phi - T\varphi, T\Phi - T\varphi) + \tfrac{1}{2}\langle \Phi - \varphi, Q(\Phi - \varphi) \rangle -$$

$$-[T(\Phi - \varphi), \sigma_T(\Phi - \varphi_B)]. \tag{3.2.16}$$

This expression can of course be obtained directly from (3.2.11). If

$$\delta^2 J \geqslant 0 \tag{3.2.17}$$

then we obtain the upper bound on $I(u, \varphi)$

$$I(u, \varphi) \leqslant J(\Phi). \tag{3.2.18}$$

Condition (3.2.17) may be satisfied in several ways, but the most obvious one is to use the fact that since Q is positive the inequality $\delta^2 J \geqslant 0$ is certainly satisfied if

$$[T(\Phi - \varphi), \sigma_T(\Phi - \varphi_B)] \leqslant 0. \tag{3.2.19}$$

This sufficient condition will be used later in applications. It can be satisfied for example by taking Φ such that $\Phi = \varphi_B$ on $\partial V_1 \subset \partial V$ and $\sigma_T^* T(\Phi - \varphi) = 0$ on $\partial V - \partial V_1$, where $[\sigma_T^* u, \varphi] = [u, \sigma_T \varphi]$.

Now we turn to the complementary principle involving the functional $G(U)$. By (2.5.37) G is defined as

$$G(U) = I(U, \Theta), \qquad (3.2.20)$$

where it is assumed that the second canonical equation $T^*U = \delta W/\delta\Phi$ has the solution

$$\Phi = \Theta(U). \qquad (3.2.21)$$

From (3.2.2) this means that

$$\Theta(U) = Q_l^{-1}(f - T^*U) \qquad (Q \neq 0), \qquad (3.2.22)$$

assuming that Q has a left inverse Q_l^{-1} such that

$$Q_l^{-1}Q = 1. \qquad (3.2.23)$$

Note that we do not require Q to have an inverse Q^{-1} such that $Q^{-1}Q = 1 = QQ^{-1}$. In fact Q^{-1} need not exist, as for example in the case when Q is an integral operator. Using (3.2.22), we therefore have

$$G(U) = I(U, Q_l^{-1}(f - T^*U)), \qquad (3.2.24)$$

and using (3.2.5) we find that

$$G(U) = -\tfrac{1}{2}(U, U) - \tfrac{1}{2}\langle f - T^*U, Q_l^{-1}(f - T^*U)\rangle + [U, \sigma_T\varphi_B]$$
$$(Q \neq 0). \quad (3.2.25)$$

This expression for $G(U)$ is quadratic as expected, and so expansion about the exact function u gives

$$G(U) = I(u, \varphi) + \delta^2 G(U), \qquad (3.2.26)$$

where from (2.5.39)

$$\delta^2 G = -\frac{1}{2}\left\{ \left(U - u, \frac{\delta^2 W}{\delta u^2}(U - u)\right) - \left\langle \Theta - \varphi, \frac{\delta^2 W}{\delta\varphi^2}(\Theta - \varphi)\right\rangle\right\}$$
$$= -\tfrac{1}{2}\{(U - u, U - u) + \langle \Theta - \varphi, Q(\Theta - \varphi)\rangle\} \qquad (3.2.27)$$

by (3.2.15). Now Q is positive and so we have immediately

$$\delta^2 G \leqslant 0 \qquad (3.2.28)$$

for all trial functions U in D_{T*}. It follows from (3.2.26) that

$$G(U) \leqslant I(u, \varphi), \qquad (3.2.29)$$

which is the complementary bound to that in (3.2.18).

If $Q = 0$ the argument involving $J(\Phi)$ goes through as before, but that for $G(U)$ must be modified. The modification comes at the introduction of $\Theta(U)$ in (3.2.21) because when $Q = 0$ the equation

$T^*U = \delta W/\delta \Phi$ becomes $T^*U = f$ and there is no unique solution $\Phi = \Theta(U)$ of this equation (cf. equation (1.4.22)). What this means is that U must be chosen to satisfy the *constraint*

$$T^*U = f \quad \text{in} \quad V. \tag{3.2.30}$$

Then

$$G(U) = I(U, \Phi) = -\tfrac{1}{2}(U, U) + [U, \sigma_T \varphi_B] \tag{3.2.31}$$

by (3.2.5), and

$$\delta^2 G = -\tfrac{1}{2}(U-u, U-u) \leqslant 0, \tag{3.2.32}$$

so that

$$G(U) \leqslant I(u, \varphi) \tag{3.2.33}$$

as before.

Since the exact function u is related to φ by $u = T\varphi$ according to (3.2.1), it is convenient to suppose that the trial function U has the form

$$U = T\Psi, \tag{3.2.34}$$

where $\Psi \in D_T$ is an approximation to φ. We then find from (3.2.25) and (3.2.31) that

$$G(T\Psi) = -\tfrac{1}{2}(T\Psi, T\Psi) - \tfrac{1}{2}\langle(f - T^*T\Psi), Q_i^{-1}(f - T^*T\Psi)\rangle +$$
$$+ [T\Psi, \sigma_T \varphi_B] \qquad (Q \neq 0), \quad (3.2.35)$$

$$= -\tfrac{1}{2}(T\Psi, T\Psi) + [T\Psi, \sigma_T \varphi_B] \qquad (Q = 0, T^*T\Psi = f \text{ in } V). \tag{3.2.36}$$

The first term on the right hand side can be rewritten with the help of (3.1.4), and we obtain

$$G(T\Psi) = -\tfrac{1}{2}\langle\Psi, T^*T\Psi\rangle - \tfrac{1}{2}\langle(f - T^*T\Psi), Q_i^{-1}(f - T^*T\Psi)\rangle -$$
$$- [T\Psi, \sigma_T(\tfrac{1}{2}\Psi - \varphi_B)] \qquad (Q \neq 0), \quad (3.2.37)$$

$$= -\tfrac{1}{2}\langle\Psi, T^*T\Psi\rangle - [T\Psi, \sigma_T(\tfrac{1}{2}\Psi - \varphi_B)]$$
$$(Q = 0, T^*T\Psi = f \text{ in } V). \quad (3.2.38)$$

The advantage of these expressions is that, when there are no boundary terms present, as for example in the case $\sigma_T = 0$, knowledge of the individual operators T and T^* is not required.

An alternative form of $G(T\Psi)$ which relates it to the functional J can be obtained after some manipulation. It is a 'least squares' expression given by

$$G(T\Psi) = J(\Psi) - \tfrac{1}{2}\langle f - A\Psi, Q_i^{-1}(f - A\Psi)\rangle +$$
$$+ \tfrac{1}{2}\langle\Psi, (1 - QQ_i^{-1})(f - T^*T\Psi)\rangle \qquad Q \neq 0, \quad (3.2.39)$$

$$= J(\Psi) \qquad Q = 0. \tag{3.2.40}$$

These are the basic results for the class of linear problems in (3.1.1)–(3.1.3), and for reference purposes it is convenient to summarize them here.

SUMMARY

1. *Linear problem*

$$A\varphi = (T^*T + Q)\varphi = f \quad \text{in} \quad V, \tag{3.2.41}$$

$$\sigma_T(\varphi - \varphi_B) = 0 \quad \text{on} \quad \partial V, \tag{3.2.42}$$

where Q is a positive symmetric operator, T^* is the adjoint of T defined by (3.1.4)

$$(u, T\varphi) = \langle T^*u, \varphi\rangle + [u, \sigma_T\varphi], \tag{3.2.43}$$

f is a known function, and φ_B prescribes the value of the exact solution φ on the boundary ∂V of some region V.

2. *Associated variational problem*

The basic functionals are

$$J(\Phi) = \tfrac{1}{2}(T\Phi, T\Phi) + \tfrac{1}{2}\langle\Phi, Q\Phi\rangle - \langle f, \Phi\rangle - [T\Phi, \sigma_T(\Phi - \varphi_B)],$$
$$\tag{3.2.44}$$

$$= \tfrac{1}{2}\langle\Phi, A\Phi\rangle - \langle f, \Phi\rangle - [T\Phi, \sigma_T(\tfrac{1}{2}\Phi - \varphi_B)]. \tag{3.2.45}$$

$$\left.\begin{aligned} G(T\Psi) &= -\tfrac{1}{2}(T\Psi, T\Psi) - \tfrac{1}{2}\langle(f - T^*T\Psi), Q_l^{-1}(f - T^*T\Psi)\rangle + \\ &\qquad\qquad + [T\Psi, \sigma_T\varphi_B], \\ &= -\tfrac{1}{2}\langle\Psi, T^*T\Psi\rangle - \tfrac{1}{2}\langle(f - T^*T\Psi), Q_l^{-1}(f - T^*T\Psi)\rangle - \\ &\qquad\qquad - [T\Psi, \sigma_T(\tfrac{1}{2}\Psi - \varphi_B)] \end{aligned}\right\} Q \neq 0.$$
$$\tag{3.2.46}$$

$$\left.\begin{aligned} &= -\tfrac{1}{2}(T\Psi, T\Psi) + \lfloor T\Psi, \sigma_T\varphi_B] \\ &= -\tfrac{1}{2}\langle\Psi, T^*T\Psi\rangle - [T\Psi, \sigma_T(\tfrac{1}{2}\Psi - \varphi_B)] \end{aligned}\right\} \quad Q = 0. \tag{3.2.47}$$

$$I(\varphi) \equiv I(u, \varphi) = -\tfrac{1}{2}\langle f, \varphi\rangle + \tfrac{1}{2}[T\varphi, \sigma_T\varphi_B]. \tag{3.2.48}$$

3. *Extremum principles:*

Minimum principle $\qquad I(\varphi) \leqslant J(\Phi) \qquad\qquad\qquad (3.2.49)$

if condition (3.2.19), namely $[T(\Phi - \varphi), \sigma_T(\Phi - \varphi_B)] \leqslant 0$, is satisfied.

Maximum principle $\qquad G(T\Psi) \leqslant I(\varphi) \qquad\qquad\qquad (3.2.50)$

for any $\Psi \in D_T$ if $Q \neq 0$, and for $\Psi \in \{\Psi : T^*T\Psi = f \text{ in } V\}$ if $Q = 0$.

3.3. Rayleigh–Ritz method

As we have just seen, the solution φ of the boundary-value problem (3.2.41), (3.2.42) can be characterized as the function which yields a minimum value to the functional $J(\Phi)$ and a maximum value to the functional $G(T\Psi)$, within suitably defined admissible classes of functions. These maximum and minimum principles provide a way of constructing approximate solutions Φ and Ψ of the problem in (3.2.41), (3.2.42) and, of course, provide upper and lower bounds for the quantity $I(\varphi)$ which is often of considerable interest. The principal method for constructing Φ and Ψ is the Rayleigh–Ritz method [67, 68], in which one minimizes $J(\Phi)$, or maximizes $G(T\Psi)$, not for the complete set of admissible functions, but only within a smaller set of functions R_n, where R_n is the n-dimensional space of linear combinations of n independent admissible functions $\Phi_1, \ldots, \Phi_n$. Then

$$I(\varphi) = \min_{\Phi \in D_A} J(\Phi) \leqslant \min_{\Phi \in R_n} J(\Phi), \tag{3.3.1}$$

and

$$\max_{\Psi \in R_n} G(T\Psi) \leqslant \max_{\Psi \in D_A} G(T\Psi) = I(\varphi). \tag{3.3.2}$$

The idea is therefore to evaluate the right-hand side of (3.3.1) and the left-hand side of (3.3.2) exactly, and hence obtain upper and lower bounds to $I(\varphi)$. At the same time, the functions Φ and Ψ, which give the minimum and maximum respectively on R_n, can be considered as approximations to the exact function φ.

Every function Φ in R_n can be written

$$\Phi = \sum_{i=1}^{n} \alpha_i \Phi_i, \tag{3.3.3}$$

where $\alpha_1, \ldots, \alpha_n$ are real constants. Similarly we can write Ψ in R_n as

$$\Psi = \sum_{i=1}^{n} \beta_i \Phi_i. \tag{3.3.4}$$

The constants $\{\alpha_i\}$ are determined by minimizing $J(\Phi)$, that is, by solving

$$\frac{\partial}{\partial \alpha_j} J\left(\sum_{i=1}^{n} \alpha_i \Phi_i \right) = 0 \qquad j = 1, 2, \ldots, n, \tag{3.3.5}$$

and the constants $\{\beta_i\}$ are determined by maximizing $G(T\Psi)$, which leads to

$$\frac{\partial}{\partial \beta_j} G\left(T \sum_{i=1}^{n} \beta_i \Phi_i \right) = 0 \qquad j = 1, 2, \ldots, n. \tag{3.3.6}$$

In this way we obtain the Rayleigh–Ritz approximations Φ and Ψ, and the corresponding bounds $J(\Phi)$ and $G(T\Psi)$ for $I(\varphi)$.

In the simplest case corresponding to $n = 1$, the Rayleigh–Ritz procedure leads to scale-independent expressions for the basic functionals. Thus, if $\tilde{\Phi}$ denotes the function in R_1, we can take

$$\Phi = \alpha \tilde{\Phi}, \tag{3.3.7}$$

and determine the constant α from the condition

$$\frac{\partial}{\partial \alpha} J(\alpha \tilde{\Phi}) = 0. \tag{3.3.8}$$

Similarly we can take $\Psi = \beta \tilde{\Phi},$ (3.3.9)

where β is determined by

$$\frac{\partial}{\partial \beta} G(\beta T \tilde{\Phi}) = 0. \tag{3.3.10}$$

Using (3.2.45) and (3.2.46) we find that the corresponding optimum values are

$$J(\alpha \tilde{\Phi}) = -\frac{1}{2} \frac{\{\langle f, \tilde{\Phi} \rangle - [T\tilde{\Phi}, \sigma_T \varphi_B]\}^2}{\langle \tilde{\Phi}, A\tilde{\Phi} \rangle - [T\tilde{\Phi}, \sigma_T \tilde{\Phi}]}, \tag{3.3.11}$$

obtained at $\alpha = \dfrac{\langle f, \tilde{\Phi} \rangle - [T\tilde{\Phi}, \sigma_T \varphi_B]}{\langle \tilde{\Phi}, A\tilde{\Phi} \rangle - [T\tilde{\Phi}, \sigma_T \tilde{\Phi}]},$ (3.3.12)

and

$$G(T\beta\tilde{\Phi}) = \frac{1}{2} \frac{\{\langle f, Q_l^{-1} A\tilde{\Phi} - \tilde{\Phi} \rangle + [T\tilde{\Phi}, \sigma_T \varphi_B]\}^2}{\langle A\tilde{\Phi}, Q_l^{-1} A\tilde{\Phi} - \tilde{\Phi} \rangle + [T\tilde{\Phi}, \sigma_T \tilde{\Phi}]} - \tfrac{1}{2}\langle f, Q_l^{-1} f \rangle \quad (Q \neq 0), \tag{3.3.13}$$

obtained at $\beta = \dfrac{\langle f, Q_l^{-1} A\tilde{\Phi} - \tilde{\Phi} \rangle + [T\tilde{\Phi}, \sigma_T \varphi_B]}{\langle A\tilde{\Phi}, Q_l^{-1} A\tilde{\Phi} - \tilde{\Phi} \rangle + [T\tilde{\Phi}, \sigma_T \tilde{\Phi}]}.$ (3.3.14)

When $J(\Phi)$ is taken in the scale-independent form (3.3.11), the result $I(\varphi) \leqslant J(\Phi)$ of (3.2.49) is usually known as the Schwinger–Levine [48] principle. The complementary principle is then given by using the expression (3.3.13) in the result $G(T\Psi) \leqslant I(\Phi)$ of (3.2.50).

3.4. The Ritz and Temple bounds

Upper and lower bounds for the lowest eigenvalue of a symmetric operator which is bounded below have been known for a considerable time [68, 79]. In this section we show how they can be derived from the general theory of this chapter [cf. 70].

Consider the eigenvalue problem

$$h\varphi = \epsilon_0 \varphi, \tag{3.4.1}$$

where h is symmetric and bounded below with eigenvalues

$$\epsilon_0 < \epsilon_1 < \epsilon_2 < \cdots,$$

and φ vanishes on ∂V. This problem can be written in the notation of section 3.2 by choosing

$$T^*T = h - \epsilon_0, \qquad Q = f = 0, \qquad \varphi_B = 0. \tag{3.4.2}$$

Then by (3.2.49) we find that the $J(\Phi)$ functional gives

$$\langle \Phi, (h-\epsilon_0)\Phi \rangle \geqslant 0, \tag{3.4.3}$$

where Φ vanishes on ∂V. If Φ is normalized so that

$$\langle \Phi, \Phi \rangle = 1, \tag{3.4.4}$$

equation (3.4.3) becomes
$$\epsilon_0 \leqslant \langle \Phi, h\Phi \rangle. \tag{3.4.5}$$

This is the Ritz [68] upper bound to the lowest eigenvalue ϵ_0.

A lower bound to ϵ_0 can be derived by using a slightly different decomposition. This involves writing

$$h-\epsilon_0 = (h-\epsilon_1)+(\epsilon_1-\epsilon_0), \tag{3.4.6}$$

taking

$$T^*T = h-\epsilon_1, \quad Q = \epsilon_1-\epsilon_0 > 0, \quad f = 0, \quad \varphi_B = 0, \tag{3.4.7}$$

and assuming that functions in the domain of T^*T are orthogonal to the eigenfunction φ of (3.4.1). Then by (3.2.50) we find that the $G(T\Psi)$ functional gives

$$-\langle \Psi, (h-\epsilon_1)\Psi \rangle - \langle (h-\epsilon_1)\Psi, (\epsilon_1-\epsilon_0)^{-1}(h-\epsilon_1)\Psi \rangle \leqslant 0, \tag{3.4.8}$$

where $\langle \Psi, \varphi \rangle = 0$ and Ψ is zero on ∂V. If Ψ is normalized, it follows from (3.4.8) that
$$\epsilon_0 \geqslant \langle \Psi, h\Psi \rangle - \frac{\langle h\Psi, h\Psi \rangle - \langle \Psi, h\Psi \rangle^2}{\epsilon_1 - \langle \Psi, h\Psi \rangle}, \tag{3.4.9}$$

provided that
$$\langle \Psi, h\Psi \rangle < \epsilon_1. \tag{3.4.10}$$

The result (3.4.9) is the Temple [79] lower bound for the lowest eigenvalue ϵ_0 of (3.4.1).

3.5. Potential theory

We now turn to the Dirichlet problem for Laplace's equation in a bounded region V with boundary ∂V. The problem is to find the solution φ of the equations
$$\nabla^2\varphi = 0 \quad \text{in} \quad V, \tag{3.5.1}$$

$$\varphi = \varphi_B \quad \text{on} \quad \partial V, \tag{3.5.2}$$

where φ_B is a real continuous function. To use the results of section 3.2 we choose
$$T = \text{grad}, \quad T^* = -\text{div}, \quad \sigma_T = \mathbf{n}, \tag{3.5.3}$$

$$Q = 0, \quad f = 0. \tag{3.5.4}$$

In this case complementary variational principles can be found. The basic functionals $J(\Phi)$ and $G(T\Psi)$ are, from (3.2.44) and (3.2.47),

$$J(\Phi) = \tfrac{1}{2}(\text{grad } \Phi, \text{grad } \Phi) - [\text{grad } \Phi, \mathbf{n}(\Phi - \varphi_B)], \qquad (3.5.5)$$

and $\qquad G(T\Psi) = -\tfrac{1}{2}(\text{grad } \Psi, \text{grad } \Psi) + [\text{grad } \Psi, \mathbf{n}\varphi_B].$ $\qquad (3.5.6)$

By (3.2.49) and (3.2.50) we therefore obtain the complementary principles

$$G(T\Psi) \leqslant I(\varphi) \leqslant J(\Phi), \qquad (3.5.7)$$

where $\qquad J(\Phi) = \tfrac{1}{2}(\text{grad } \Phi, \text{grad } \Phi), \qquad \Phi = \varphi_B \quad \text{on} \quad \partial V, \qquad (3.5.8)$

$$I(\varphi) = \tfrac{1}{2}(\text{grad } \varphi, \text{grad } \varphi), \qquad (3.5.9)$$

$G(T\Psi) = -\tfrac{1}{2}(\text{grad } \Psi, \text{grad } \Psi) + [\text{grad } \Psi, \mathbf{n}\varphi_B],$

$$\nabla^2\Psi = 0 \quad \text{in} \quad V. \quad (3.5.10)$$

An interesting application of these bounds can be made when V is the space *external* to some convex surface S and

$$\varphi_B = 1 \text{ on } S, \qquad \varphi_B = \text{order } \frac{1}{r} \text{ at } \infty. \qquad (3.5.11)$$

In this case, the capacity C of the surface S is given by

$$C = \frac{1}{4\pi} \int_V \text{grad } \varphi \cdot \text{grad } \varphi \, dV, \qquad (3.5.12)$$

where φ is the solution of the problem in (3.5.1) and (3.5.2). Since $I(\varphi)$ in (3.5.9) is proportional to the capacity C, the result in (3.5.7) provides bounds for the capacity of a surface. These bounds are readily seen to be

$$C_-(\Psi) \leqslant C \leqslant C_+(\Phi), \qquad (3.5.13)$$

where

$$C_+(\Phi) = \frac{1}{4\pi} \int_V \text{grad } \Phi \cdot \text{grad } \Phi \, dV, \qquad \Phi = \varphi_B \quad \text{on} \quad \partial V, \quad (3.5.14)$$

and

$$C_-(\Psi) = \frac{1}{4\pi} \left\{ 2 \int_{\partial V} \varphi_B \mathbf{n} \cdot \text{grad } \Psi \, dB - \int_V \text{grad } \Psi \cdot \text{grad } \Psi \, dV \right\},$$

$$\nabla^2\Psi = 0 \quad \text{in} \quad V. \quad (3.5.15)$$

A scale-independent form of $C_-(\Psi)$ follows on taking $\Psi = \beta \tilde{\Psi}$ and determining β from $\partial C_- / \partial \beta = 0$. This gives

$$C_-(\beta\tilde{\Psi}) = \frac{1}{4\pi} \frac{\left(\int_{\partial V} \varphi_B \, \mathbf{n} \cdot \text{grad } \tilde{\Psi} \, dB \right)^2}{\int_V \text{grad } \tilde{\Psi} \cdot \text{grad } \tilde{\Psi} \, dV}, \qquad \nabla^2\tilde{\Psi} = 0 \quad \text{in} \quad V. \quad (3.5.16)$$

The upper bound (3.5.14) is the well-known Dirichlet bound which is given by standard Euler–Lagrange theory, while the lower bound (3.5.16) has been derived in the literature by means of the Schwarz inequality [cf. 77]. Use of the hypercircle method [78] would also lead to these bounds. The present derivation shows how they arise naturally from complementary variational principles [18].

Example. Bounds for the capacity of a cube.

We illustrate these theoretical results by calculating bounds for the capacity of a cube of side 2.

To obtain a lower bound we take the trial function

$$\Psi = \frac{1}{r} \tag{3.5.17}$$

in (3.5.16), which satisfies the constraint $\nabla^2\Psi = 0$ in V. We then find that

$$\int_{\partial V} \varphi_B \mathbf{n} . \operatorname{grad} \Psi \, dB = 4\pi, \tag{3.5.18}$$

and

$$\int_V \operatorname{grad} \Psi . \operatorname{grad} \Psi \, dV = 8\sqrt{2} \left\{ \frac{\pi}{2} - \tan^{-1}\sqrt{2} + 2\tan^{-1}\frac{1}{\sqrt{2}} \right\}, \tag{3.5.19}$$

which lead to the lower bound

$$C_- = 1 \cdot 202. \tag{3.5.20}$$

A slightly better lower bound can be obtained from the use of the volume radius [60]. The result is

$$C_- = 1 \cdot 241. \tag{3.5.21}$$

Turning now to the upper bound (3.5.14), we find that, if the trial function Φ is symmetric with respect to the cube and each of the four triangles obtained by joining the diagonals in a face, then

$$C_+(\Phi) = \frac{24}{4\pi} \int_1^\infty dx \int_0^x dz \int_{-z}^z (\operatorname{grad} \Phi)^2 \, dy. \tag{3.5.22}$$

A very simple trial function is

$$\Phi(x) = \frac{1}{x} \tag{3.5.23}$$

which satisfies $\Phi = 1$ on the face $x = 1$, and leads to

$$C_+ = 6/\pi = 1 \cdot 910. \tag{3.5.24}$$

This upper-bound result has been improved on by Conlan, Diaz, and Parr [27], who chose a more elaborate trial function

$$\Phi = \frac{1}{x} + (\lambda \, |y| + \mu z)\left(\frac{1}{x^3} - \frac{1}{x^2}\right), \qquad (3.5.25)$$

where λ and μ were determined by $\partial C_+/\partial \lambda = 0$, $\partial C_+/\partial \mu = 0$. This gives the bound

$$C_+ = 1{\cdot}6103. \qquad (3.5.26)$$

3.6. Electrostatics

The results in section 3.5 can be extended slightly by considering the electrostatic field equation

$$-\nabla^2 \varphi = 4\pi\rho \quad \text{in} \quad V, \qquad (3.6.1)$$

$$\varphi = \varphi_B \quad \text{on} \quad \partial V. \qquad (3.6.2)$$

Here φ is the electrostatic potential and ρ is the charge density. This problem corresponds to the choice

$$T = \text{grad}, \qquad T^* = -\text{div}, \qquad \sigma_T = \mathbf{n}, \qquad (3.6.3)$$

$$Q = 0, \qquad f = 4\pi\rho. \qquad (3.6.4)$$

Complementary variational principles can be found for such a problem and the basic functionals $J(\Phi)$ and $G(T\Psi)$ are, from (3.2.44) and (3.2.47),

$$J(\Phi) = \tfrac{1}{2}(\text{grad } \Phi, \text{grad } \Phi) - \langle 4\pi\rho, \Phi \rangle - [\text{grad } \Phi, \mathbf{n}(\Phi - \varphi_B)], \quad (3.6.5)$$

and $\qquad G(T\Psi) = -\tfrac{1}{2}(\text{grad } \Psi, \text{grad } \Psi) + [\text{grad } \Psi, \mathbf{n}\varphi_B]. \qquad (3.6.6)$

By (3.2.44) and (3.2.50) we therefore obtain the complementary principles

$$G(T\Psi) \leqslant I(\varphi) \leqslant J(\Phi), \qquad (3.6.7)$$

where

$$J(\Phi) = \tfrac{1}{2}(\text{grad } \Phi, \text{grad } \Phi) - \langle 4\pi\rho, \Phi \rangle, \qquad \Phi = \varphi_B \quad \text{on} \quad \partial V, \quad (3.6.8)$$

$$I(\varphi) = \tfrac{1}{2}(\text{grad } \varphi, \text{grad } \varphi) - \langle 4\pi\rho, \varphi \rangle, \qquad (3.6.9)$$

$$G(T\Psi) = -\tfrac{1}{2}(\text{grad } \Psi, \text{grad } \Psi) + [\text{grad } \Psi, \mathbf{n}\varphi_B],$$

$$-\nabla^2\Psi = 4\pi\rho \quad \text{in} \quad V. \qquad (3.6.10)$$

The bounds in (3.6.7) correspond to the Dirichlet and Thomson bounds in electrostatics [cf. 28]. Here we have shown how they follow directly from the results of section 3.2 [15, 18].

3.7. Magnetostatics

The magnetostatic analogue of the problem discussed in the previous section is described by the equation

$$\text{curl curl } \boldsymbol{\varphi} = 4\pi\mathbf{j} \quad \text{in} \quad V, \tag{3.7.1}$$

with
$$\boldsymbol{\varphi} = \boldsymbol{\varphi}_B \quad \text{on} \quad \partial V, \tag{3.7.2}$$

where $\mathbf{j}$ is a steady current in a medium of unit permeability and $\boldsymbol{\varphi}$ is the vector potential (usually denoted by $\mathbf{A}$). This boundary-value problem corresponds to

$$T = \text{curl}, \qquad T^* = \text{curl}, \qquad \sigma_T = \mathbf{n}\wedge, \tag{3.7.3}$$

$$Q = 0, \qquad f = 4\pi\mathbf{j}. \tag{3.7.4}$$

From section 3.2 we see that complementary variational principles can be obtained for this problem. The basic functionals $J(\Phi)$ and $G(T\Psi)$ are, from (3.2.44) and (3.2.47),

$$J(\Phi) = \tfrac{1}{2}(\text{curl }\Phi, \text{curl }\Phi) - \langle 4\pi\mathbf{j}, \Phi\rangle - [\text{curl }\Phi, \mathbf{n} \wedge (\Phi - \boldsymbol{\varphi}_B)] \tag{3.7.5}$$

and
$$G(T\Psi) = -\tfrac{1}{2}(\text{curl }\Psi, \text{curl }\Psi) + [\text{curl }\Psi, \mathbf{n} \wedge \boldsymbol{\varphi}_B]. \tag{3.7.6}$$

By (3.2.49) and (3.2.50) we therefore obtain the complementary bounds [18]

$$G(T\Psi) \leqslant I(\varphi) \leqslant J(\Phi), \tag{3.7.7}$$

where

$$J(\Phi) = \tfrac{1}{2}(\text{curl }\Phi, \text{curl }\Phi) - \langle 4\pi\mathbf{j}, \Phi\rangle, \qquad \Phi = \boldsymbol{\varphi}_B \quad \text{on} \quad \partial V, \tag{3.7.8}$$

$$I(\boldsymbol{\varphi}) = \tfrac{1}{2}(\text{curl }\boldsymbol{\varphi}, \text{curl }\boldsymbol{\varphi}) - \langle 4\pi\mathbf{j}, \boldsymbol{\varphi}\rangle, \tag{3.7.9}$$

$$G(T\Psi) = -\tfrac{1}{2}(\text{curl }\Psi, \text{curl }\Psi) + [\text{curl }\Psi, \mathbf{n} \wedge \boldsymbol{\varphi}_B],$$

$$\text{curl curl }\Psi = 4\pi\mathbf{j} \quad \text{in} \quad V. \tag{3.7.10}$$

For $\boldsymbol{\varphi}_B = 0$ these bounds reduce to those of Schrader [73], who obtained his results by using the analogy with the Dirichlet and Thomson principles in electrostatics. In this case

$$I(\boldsymbol{\varphi}) = -2\pi\langle\mathbf{j}, \boldsymbol{\varphi}\rangle = -4\pi\Omega, \tag{3.7.11}$$

where Ω is the self-interaction energy of a steady current $\mathbf{j}$ in a medium of unit permeability. From (3.7.7) we then obtain complementary bounds for Ω

$$\Omega_-(\Phi) \leqslant \Omega \leqslant \Omega_+(\Psi), \tag{3.7.12}$$

where

$$\Omega_-(\Phi) = \langle\mathbf{j}, \Phi\rangle - \frac{1}{8\pi}(\text{curl }\Phi, \text{curl }\Phi), \qquad \Phi = 0 \quad \text{on} \quad \partial V, \tag{3.7.13}$$

and

$$\Omega_+(\Psi) = \frac{1}{8\pi}(\text{curl }\Psi, \text{curl }\Psi), \qquad \text{curl curl }\Psi = 4\pi\mathbf{j} \quad \text{in} \quad V. \quad (3.7.14)$$

Example. To illustrate these bounds for Ω we use an example given by Schrader [73] and consider a charge density

$$\rho(\mathbf{r}) = \frac{1}{4\pi}(4-r)e^{-r}\sin\theta, \qquad (3.7.15)$$

which rotates about the z-axis with unit angular velocity. The current density is

$$\mathbf{j} = \hat{\boldsymbol{\varphi}}\frac{1}{4\pi}(4-r)e^{-r}\sin\theta, \qquad (3.7.16)$$

where $\hat{\boldsymbol{\varphi}}$ denotes the usual unit vector in spherical polar coordinates. It is readily shown that suitable solutions for the vector potential $\boldsymbol{\varphi}$ and magnetic field $\mathbf{u}$ are

$$\boldsymbol{\varphi} = \hat{\boldsymbol{\varphi}}re^{-r}\sin\theta, \qquad (3.7.17)$$

and

$$\mathbf{u} = 2\hat{\mathbf{r}}e^{-r}\cos\theta + \hat{\boldsymbol{\theta}}(r-2)e^{-r}\sin\theta, \qquad (3.7.18)$$

since these satisfy

$$\text{curl }\boldsymbol{\varphi} = \mathbf{u}, \qquad \text{curl }\mathbf{u} = 4\pi\mathbf{j}. \qquad (3.7.19)$$

Then

$$\Omega = \tfrac{1}{2}\langle\mathbf{j}, \boldsymbol{\varphi}\rangle = \frac{1}{8\pi}\int(\text{curl }\boldsymbol{\varphi})^2\,\mathrm{d}V = \tfrac{1}{4}. \qquad (3.7.20)$$

To calculate a lower bound we take the trial function

$$\boldsymbol{\Phi} = \lambda\hat{\boldsymbol{\varphi}}r^3e^{-\alpha r}\sin^2\theta \qquad (3.7.21)$$

in (3.7.13), where λ and α are variational parameters determined by $\partial\Omega_-/\partial\lambda = 0$, $\partial\Omega_-/\partial\alpha = 0$. The result is [73]

$$\Omega_- = 0{\cdot}2133 \qquad (3.7.22)$$

corresponding to $\alpha = 1{\cdot}596$.

To calculate an upper bound we take the trial function Ψ such that

$$\text{curl }\Psi = -\hat{\mathbf{r}}\cos\theta(r-4)re^{-r} \qquad (3.7.23)$$

which satisfies the constraint curl curl $\Psi = 4\pi\mathbf{j}$. Then (3.7.14) gives

$$\Omega_+ = 0{\cdot}4375. \qquad (3.7.24)$$

This value can be improved [73] by taking

$$\text{curl }\Psi = -\hat{\mathbf{r}}\cos\theta(r-4)re^{-r} + \text{grad}\{e^{-\beta r}\cos\theta\}, \qquad (3.7.25)$$

where β is a variational parameter determined by $\partial\Omega_+/\partial\beta = 0$. The result is

$$\Omega_+ = 0\cdot 3625, \qquad (3.7.26)$$

corresponding to $\beta = 1$.

3.8. Diffusion

Diffusion problems represent another class of boundary-value problems for which complementary variational principles can be obtained. To illustrate this we shall consider the case of diffusion of neutrons in a reactor with volume V. When a uniform isotropic source of unit total strength exists in V, the flux φ of neutrons is given according to diffusion theory [83] by the solution of

$$(-\nabla^2 + \kappa^2)\varphi = k \qquad \text{in} \quad V, \qquad (3.8.1)$$

with

$$\varphi = 0 \qquad \text{on} \quad \partial V, \qquad (3.8.2)$$

where $\kappa^2 = (1-C)/D$ and $k = (VD)^{-1}$. Here D is the diffusion constant and C (<1) is the average number of secondary neutrons emerging from each collision which make the reactor critical. This boundary-value problem corresponds to

$$T = \text{grad}, \qquad T^* = -\text{div}, \qquad \sigma_T = \mathbf{n}, \qquad (3.8.3)$$

$$Q = \kappa^2, \qquad f = k, \qquad \varphi_B = 0. \qquad (3.8.4)$$

Since Q here is positive we see from section 3.2 that complementary principles can be found for this problem. From (3.2.44) and (3.2.46) the basic functionals $J(\Phi)$ and $G(T\Psi)$ are

$$J(\Phi) = \tfrac{1}{2}(\text{grad } \Phi, \text{grad } \Phi) + \tfrac{1}{2}\langle \Phi, \kappa^2\Phi\rangle - \langle k,\Phi\rangle - [\text{grad } \Phi, \mathbf{n}\Phi] \quad (3.8.5)$$

and $G(T\Psi) = -\tfrac{1}{2}(\text{grad }\Psi, \text{grad }\Psi) - \tfrac{1}{2}\langle(k+\nabla^2\Psi), \kappa^{-2}(k+\nabla^2\Psi)\rangle.$ (3.8.6)

By (3.2.49) and (3.2.50) we therefore obtain the complementary bounds

$$G(T\Psi) \leqslant I(\varphi) \leqslant J(\Phi), \qquad (3.8.7)$$

where

$$J(\Phi) = \tfrac{1}{2}(\text{grad } \Phi, \text{grad } \Phi) + \tfrac{1}{2}\langle \Phi, \kappa^2\Phi\rangle - \langle k,\Phi\rangle, \quad \Phi = 0 \text{ on } \partial V, \quad (3.8.8)$$

$$I(\varphi) = J(\varphi) = -\tfrac{1}{2}\langle k,\varphi\rangle, \qquad (3.8.9)$$

$$G(T\Psi) = -\tfrac{1}{2}(\text{grad } \Psi, \text{grad } \Psi) - \tfrac{1}{2}\langle k+\nabla^2\Psi, \kappa^{-2}(k+\nabla^2\Psi)\rangle. \qquad (3.8.10)$$

Since the absorption probability for this diffusion process is defined by [cf. 32]

$$P = \langle 1-C, \varphi\rangle, \qquad (3.8.11)$$

we have from (3.8.9) that

$$P = -\frac{2}{k}(1-C)I(\varphi).\tag{3.8.12}$$

Hence, from (3.8.7)

$$P_1(\Phi) \leqslant P \leqslant P_2(\Psi),\tag{3.8.13}$$

where

$$P_1(\Phi) = -\frac{2}{k}(1-C)J(\Phi), \qquad P_2(\Psi) = -\frac{2}{k}(1-C)G(T\Psi).\tag{3.8.14}$$

Equation (3.8.13) thus provides upper and lower bounds for the absorption probability [10].

Scale-independent forms of P_1 and P_2 are readily obtained from (3.3.11) and (3.3.13) and are

$$P_1(\Phi) = \frac{\kappa^2}{V}\frac{\left(\int_V \Phi\, dV\right)^2}{\int_V \{(\mathrm{grad}\ \Phi)^2 + \kappa^2 \Phi^2\}\, dV}, \qquad \Phi = 0 \text{ on } \partial V,\tag{3.8.15}$$

and

$$P_2(\Psi) = 1 - \frac{\left(\int_V \nabla^2\Psi\, dV\right)^2}{V\int_V \{(\nabla^2\Psi)^2 + \kappa^2(\mathrm{grad}\ \Psi)^2\}\, dV}.\tag{3.8.16}$$

Equation (3.8.15) is identical to the formula for the lower bound of P which Dresner [32] derived from a generalized form of the Schwarz inequality. The upper-bound formula (3.8.16) was derived in [10] from the theory of complementary principles.

Example.   Neutrons in a sphere.

To illustrate this theory we shall evaluate upper and lower bounds for the absorption probability of neutrons in a sphere. Let a be the radius of the sphere and introduce spherical polar coordinates with the centre of the sphere as origin. We take very simple trial functions

$$\Phi = \Psi = a^2 - r^2,\tag{3.8.17}$$

which vanish on the surface of the sphere, and from (3.8.15) and (3.8.16) we obtain P_1 and P_2 as functions of κa. These bounds are compared in Table 3.1. In addition the exact value of P is readily found to be

$$P = 1 - \frac{3}{\kappa a}\left(\coth \kappa a - \frac{1}{\kappa a}\right),\tag{3.8.18}$$

which follows from (3.8.11) by using the exact neutron flux

$$\varphi = \frac{k}{\kappa^2}\left\{1 - \frac{a \sinh \kappa r}{r \sinh \kappa a}\right\}. \tag{3.8.19}$$

The exact P is also given in Table 3.1.

From the calculations we see that the upper and lower bounds given by the trial (3.8.17) are very close for small values of κa, gradually getting further apart as κa increases.

Finally, we add that an integral equation approach can also be used for problems like this [7], providing alternative upper and lower bounds for the absorption probability.

TABLE 3.1. *Probabilities P_1, P_2, and P for a sphere*

κa	0	0·1	0·5	1	5	10	50	∞
P_1	0	0·00067	0·01628	0·06087	0·49296	0·63348	0·69707	0·7
P	0	0·00067	0·01628	0·06090	0·51995	0·73000	0·94120	1
P_2	0	0·00067	0·01639	0·06250	0·62500	0·86957	0·99404	1

3.9. The Milne problem

As a second example on neutron transport we consider the Milne problem, which concerns a semi-infinite, homogeneous half-space bounded by a vacuum with a source of neutrons at infinity. The transport equation for this problem [cf. 31] may be written in the form

$$\varphi(x) = f(x) + K\varphi(x), \tag{3.9.1}$$

where

$$f(x) = \frac{c}{2\nu}[E_1(x) - e^{\nu x}E_1\{(1+\nu)x\}], \tag{3.9.2}$$

and

$$K\varphi(x) = \frac{c}{2}\int_0^\infty E_1(|x-t|)\varphi(t)\,\mathrm{d}t. \tag{3.9.3}$$

Here

$$E_1(x) \equiv \int_1^\infty \mathrm{e}^{-xt}\,\mathrm{d}t/t, \tag{3.9.4}$$

ν is determined by

$$2\nu = c\ln\{(1+\nu)/(1-\nu)\}, \tag{3.9.5}$$

and c is a real number such that

$$0 < c < 1. \tag{3.9.6}$$

The integral operator K is symmetric and positive-definite with eigenvalues $\{\lambda_i\}$ such that $\lambda_i > 0$ and the smallest eigenvalue is $\lambda_0 = 1/c$ [cf. 61]. This means that $(c-K)$ is positive and we now choose

$$T^*T = c-K, \qquad \sigma_T = 0 \tag{3.9.7}$$

$$Q = 1-c. \tag{3.9.8}$$

Since $0 < c < 1$, Q is positive, and we see from section 3.2 that complementary variational principles can be found for this problem. From (3.2.45) and (3.2.46) the basic functionals $J(\Phi)$ and $G(T\Psi)$ are

and
$$J(\Phi) = \tfrac{1}{2}\langle\Phi,(1-K)\Phi\rangle - \langle f, \Phi\rangle, \tag{3.9.9}$$

$$G(T\Psi) = -\tfrac{1}{2}\langle\Psi,(c-K)\Psi\rangle - \frac{1}{2(1-c)}\langle f-(c-K)\Psi, f-(c-K)\Psi\rangle. \tag{3.9.10}$$

Also
$$I(\varphi) = J(\varphi) = -\tfrac{1}{2}\langle f, \varphi\rangle. \tag{3.9.11}$$

The conditions for complementary principles are already satisfied here, and so from (3.2.49) and (3.2.50) we obtain

$$G(T\Psi) \leqslant I(\varphi) \leqslant J(\Phi). \tag{3.9.12}$$

If we define a quantity x_0, known as the extrapolated end-point, by

$$x_0 = -\frac{1}{2\nu}\ln\Gamma, \tag{3.9.13}$$

where
$$\Gamma = \left\{\frac{c}{2\nu}\left[\frac{1}{\nu}\ln\left(\frac{1}{1-\nu}\right)-\frac{1}{c}\right] - \langle f, \varphi\rangle\right\}\left\{\frac{c}{\nu}\left[\frac{1}{1-\nu}-\frac{1}{c}\right]\right\}^{-1}, \tag{3.9.14}$$

we see that (3.9.12) will provide complementary bounds for x_0.

With trial functions

$$\Phi = \alpha e^{-\nu x}, \qquad \Psi = \beta e^{-\nu x}, \tag{3.9.15}$$

where α and β are determined by $\partial J/\partial\alpha = 0$ and $\partial G/\partial\beta = 0$, Pomraning [61] has computed upper and lower bounds for cx_0. The results are given in Table 3.2 along with results of Case, de Hoffman, and Placzek [25], and Mark [51], obtained by a numerical computation of the exact relation

$$x_0 = \frac{c}{2\nu}\int_0^1 dt\left\{1+\frac{ct^2}{1-t^2}\right\}\frac{\tanh^{-1}(\nu t)}{[(1-ct\tanh^{-1}t)^2+(\pi ct/2)^2]}. \tag{3.9.16}$$

We see from Table 3.2 that the bounds are very close over the whole range of c. For small values of c we also note that some of the results based on a numerical integration of eqn (3.9.16) fall outside the variational bounds. The reason for this is that for very small values of c the numerical integration is very difficult to perform. By using extremely accurate methods, Pomraning and Lathrop [63] have recomputed cx_0 from (3.9.16) for small values of c and the results now fall inside the bounds given in Table 3.2.

TABLE 3.2. cx_0 *for the Milne problem*

c	Lower Bound	Case *et al.*	Mark	Upper Bound
0·1	0·853826	0·8539	0·8590	0·853829
0·2	0·78476	0·7851	0·7843	0·78479
0·3	0·74836	0·7491	0·7486	0·74843
0·4	0·7298	0·7305	0·7300	0·7300
0·5	0·7202	0·7207	0·7204	0·7206
0.6	0·7150	0·7155	0·7154	0·7156
0·7	0·7121	0·7127	0·7126	0·7132
0·8	0·7104	0·7113	0·7112	0·7126
0·9	0·7094	0·7106	0·7106	0·7148

3.10. Kirkwood–Riseman integral equation

In the Kirkwood–Riseman theory of intrinsic viscosities and diffusion constants of flexible macromolecules [43], the following singular Fredholm integral equation arises

$$\varphi(x) = f(x) + \lambda \int_{-1}^{1} |x-t|^{-\alpha} \varphi(t) \, dt, \qquad (3.10.1)$$

where $0 < \alpha < 1$. The parameter λ is negative in this theory and α is a statistical parameter which assumes the value of $\frac{1}{2}$ when a Gaussian model is employed. The kernel of this integral operator K is clearly symmetric and it can be shown [19] that

$$|x-t|^{-\alpha} = \sum_{n} a_n \theta_n(x) \theta_n(t), \qquad (3.10.2)$$

where $a_n \geqslant 0$ for all n and $\{\theta_n\}$ is a set of Gegenbauer polynomials. This means that the associated integral operator K is positive-definite.

We rewrite (3.10.1) as

$$(-\lambda K + 1)\varphi = f. \qquad (3.10.3)$$

Then we choose, remembering that λ is negative,

$$T^*T = -\lambda K, \qquad \sigma_T = 0, \qquad (3.10.4)$$

$$Q = 1, \qquad (3.10.5)$$

TABLE 3.3. *Variational parameters and bounds:* $f = 1$

(a) $\alpha = 0.2$

$-\lambda$	c_1	c_2	$-2J$	d_1	d_2	$-2G$	$(G-J)/G$
2	0·06322	0·15079	0·34372	0·06070	0·15160	0·34378	0·00017
4	0·05227	0·07675	0·18834	0·04981	0·07754	0·18843	0·00046
8	0·03685	0·03722	0·99004(−1)	0·03485	0·03786	0·99132(−1)	0·00129
16	0·02281	0·01781	0·50834(−1)	0·02145	0·01825	0·51002(−1)	0·00329
32	0·01289	0·00859	0·25772(−1)	0·01207	0·00885	0·25966(−1)	0·00747
64	0·00689	0·00419	0·12977(−1)	0·00643	0·00434	0·13189(−1)	0·01610
128	0·00356	0·00207	0·65118(−2)	0·00332	0·00215	0·67342(−2)	0·03303
256	0·00181	0·00103	0·32618(−2)	0·00169	0·00107	0·34894(−2)	0·06523
512	0·00091	0·00051	0·16324(−2)	0·00085	0·00053	0·18628(−2)	0·12370
1024	0·00046	0·00026	0·81656(−3)	0·00043	0·00027	0·10484(−2)	0·22112

Here $N(-m)$ means $N \times 10^{-m}$

(b) $\alpha = 0.5$

$-\lambda$	c_1	c_2	$-2J$	d_1	d_2	$-2G$	$(G-J)/G$
2	0·05529	0·09956	0·23598	0·05227	0·10047	0·23660	0·00262
4	0·03441	0·05131	0·12557	0·03241	0·05192	0·12636	0·00627
8	0·01943	0·02596	0·64880(−1)	0·01826	0·02632	0·65794(−1)	0·01389
16	0·01036	0·01304	0·32994(−1)	0·00972	0·01324	0·33978(−1)	0·02896
32	0·00535	0·00653	0·16639(−1)	0·00502	0·00664	0·17663(−1)	0·05798
64	0·00272	0·00327	0·83556(−2)	0·00255	0·00332	0·94004(−2)	0·11114
128	0·00137	0·00164	0·41868(−2)	0·00129	0·00166	0·52424(−2)	0·20136
256	0·00069	0·00082	0·20956(−2)	0·00065	0·00083	0·31566(−2)	0·33612
512	0·00035	0·00041	0·10484(−2)	0·00032	0·00042	0·21120(−2)	0·50379
1024	0·00017	0·00020	0·52436(−3)	0·00016	0·00021	0·15893(−2)	0·96701

(c) $\alpha = 0.8$

$-\lambda$	c_1	c_2	$-2J$	d_1	d_2	$-2G$	$(G-J)/G$
2	0·01171	0·04591	0·99618(−1)	0·01148	0·04598	0·99698(−1)	0·00080
4	0·00623	0·02347	0·51088(−1)	0·00611	0·02350	0·51178(−1)	0·00176
8	0·00321	0·01187	0·25876(−1)	0·00315	0·01189	0·25972(−1)	0·00367
16	0·00163	0·00597	0·13023(−1)	0·00160	0·00598	0·13122(−1)	0·00757
32	0·00082	0·00299	0·65328(−2)	0·00081	0·00300	0·66338(−2)	0·01523
64	0·00041	0·00150	0·32718(−2)	0·00041	0·00150	0·33736(−2)	0·03018
128	0·00021	0·00075	0·16372(−2)	0·00020	0·00075	0·17394(−2)	0·05877
256	0·00010	0·00037	0·81894(−3)	0·00010	0·00038	0·92138(−3)	0·11118
512	0·00005	0·00019	0·40956(−3)	0·00005	0·00019	0·51208(−3)	0·20020
1024	0·00003	0·00009	0·20480(−3)	0·00003	0·00009	0·30738(−3)	0·33372

TABLE 3.4. *Variational parameters and bounds:* $f = x^2$

(a) $\alpha = 0.2$

$-\lambda$	c_1	c_2	$-2J$	d_1	d_2	$-2G$	$(G-J)/G$
0·1	0·98363	-0·05753	0·35510	0·98246	-0·05728	0·35512	0·00006
0·5	0·91128	-0·14787	0·26594	0·90668	-0·14659	0·26618	0·00090
1·0	0·82827	-0·17270	0·21618	0·82101	-0·17053	0·21676	0·00268
2·0	0·69756	-0·16961	0·16595	0·68763	-0·16654	0·16721	0·00752
4·0	0·52877	-0·14022	0·11803	0·51767	-0·13672	0·12047	0·02025
8·0	0·35590	-0·09886	0·07645	0·34603	-0·09572	0·08049	0·05017

(b) $\alpha = 0.5$

$-\lambda$	c_1	c_2	$-2J$	d_1	d_2	$-2G$	$(G-J)/G$
0·1	0·90868	-0·05610	0·32608	0·90548	-0·05530	0·32618	0·00031
0·5	0·65125	-0·09346	0·19819	0·64379	-0·09141	0·19934	0·00574
1·0	0·47755	-0·08213	0·13627	0·46983	-0·07995	0·13850	0·01608
2·0	0·31069	-0·05932	0·84730(-1)	0·30432	-0·05749	0·88284(-1)	0·04026
4·0	0·18270	-0·03691	0·48472(-1)	0·17838	-0·03566	0·53206(-1)	0·08897
8·0	0·10014	-0·02084	0·26164(-1)	0·09757	-0·02009	0·31736(-1)	0·17557

(c) $\alpha = 0.8$

$-\lambda$	c_1	c_2	$-2J$	d_1	d_2	$-2G$	$(G-J)/G$
0·1	0·62552	−0·03356	0·22784	0·62443	−0·03343	0·22808	0·00105
0·5	0·24862	−0·02215	0·84682(−1)	0·24781	−0·02202	0·85544(−1)	0·01008
1·0	0·14169	−0·01376	0·47504(−1)	0·14118	−0·01368	0·48588(−1)	0·02231
2·0	0·07616	−0·00774	0·25302(−1)	0·07587	−0·00770	0·26532(−1)	0·04636
4·0	0·03956	−0·00412	0·13079(−1)	0·03940	−0·00409	0·14393(−1)	0·09130
8·0	0·02017	−0·00213	0·66520(−2)	0·02009	−0·00211	0·80116(−2)	0·16970

in the general theory of section 3.2. Since Q is positive we know that complementary principles can be found [cf. 71]. From (3.2.45) and (3.2.46), the basic functionals for this case are

$$J(\Phi) = \tfrac{1}{2}\langle \Phi, (1-\lambda K)\Phi\rangle - \langle f, \Phi\rangle, \qquad (3.10.6)$$

$$G(T\Psi) = \tfrac{1}{2}\langle \Psi, \lambda K\Psi\rangle - \tfrac{1}{2}\langle f+\lambda K\Psi, f+\lambda K\Psi\rangle, \qquad (3.10.7)$$

$$I(\varphi) = J(\varphi) = -\tfrac{1}{2}\langle f, \varphi\rangle. \qquad (3.10.8)$$

The complementary variational principles are, by (3.2.49) and (3.2.50), given by

$$G(T\Psi) \leqslant I(\varphi) \leqslant J(\Phi). \qquad (3.10.9)$$

For the trial functions Φ and Ψ in (3.10.9) we shall choose

$$\Phi = c_1 x^2 + c_2, \qquad \Psi = d_1 x^2 + d_2, \qquad (3.10.10)$$

where the variational parameters are obtained from the stationary conditions

$$\frac{\partial J}{\partial c_i} = 0, \qquad \frac{\partial G}{\partial d_i} = 0 \qquad i = 1,2. \qquad (3.10.11)$$

Calculations for a range of values of λ and α have been carried out [6] in the two cases (i) $f(x) = 1$, (ii) $f(x) = x^2$, and the results are given in Tables 3.3 and 3.4.

When $f(x) = 1$ we see from (3.10.8) that

$$I(\varphi) = -\tfrac{1}{2}\langle 1, \varphi\rangle, \qquad (3.10.12)$$

and this, by the Kirkwood–Riseman theory [43], is inversely proportional to the translational diffusion constant D, that is

$$\langle 1, \varphi\rangle = \frac{\nu}{D}, \qquad (3.10.13)$$

where $\nu = 2M_0 kT/M\zeta$, in which kT is the Boltzmann energy, ζ is a friction constant characteristic of the fluid, M is the molecular weight of the polymer unit, and M_0 is the molecular weight of the monomer unit. Since G and J provide complementary bounds for $I(\varphi)$, it follows that $-\nu/2J$ and $-\nu/2G$ obtained from Table 3.3 provide upper and lower bounds for D, the translational diffusion constant. Thus

$$-\frac{\nu}{2G} \leqslant D \leqslant -\frac{\nu}{2J} \qquad (f = 1). \qquad (3.10.14)$$

The case $\alpha = 0.5$ corresponds to a Gaussian model for the statistics of the polymer chain. It is to be expected that statistics other than

Gaussian will lead to values of α lying somewhere between zero and unity. The complementary bounds used here give results which provide a comparison between the solutions for Gaussian and non-Gaussian models.

Judging by the closeness of the bounds G and J, and by the closeness of the parameters c_1, d_1 and c_2, d_2, we see from Tables 3.3 and 3.4 that the solutions are quite accurate for the smaller λ-values shown. With these values in the case (i) $f(x) = 1$, we are able to place close upper and lower limits on values of the diffusion constant D. For fixed values of λ in the range -2 to -256, we see from Table 3.3 and (3.10.14) that the diffusion constant D increases as α is increased.

In the case (ii) $f(x) = x^2$, some comparison with other work is possible. Table 3.5 contains the numerical solution of Ullman and Ullman [81] for $\alpha = 0{\cdot}5$, $\lambda = -0{\cdot}5$, and the variational solutions Φ and Ψ, which for this case are given by

$$\Phi = (0{\cdot}65125)x^2 - 0{\cdot}09346, \qquad (3.10.15)$$
$$\Psi = (0{\cdot}64379)x^2 - 0{\cdot}09141. \qquad (3.10.16)$$

While these variational solutions are accurate to $0{\cdot}57$ per cent in terms of the bounds G and J, we see from Table 3.5 that the agreement with the numerical solution is only moderately good. This, however, must be attributed to the extreme simplicity of the trial functions in (3.10.10).

TABLE 3.5. *Comparison with results of Ullman and Ullman* [81] *for $f(x) = x^2$, $\alpha = 0{\cdot}5$, $\lambda = -0{\cdot}5$*

x	φ (ref. 81)	Φ	Ψ
0·019511	−0·081022	−0·09321	−0·09116
0·116084	−0·073681	−0·08468	−0·08273
0·227786	−0·052520	−0·05967	−0·05801
0·413779	0·015277	0·01804	0·01882
0·502804	0·062386	0·07118	0·07135
0·636054	0·15394	0·17001	0·16904
0·778306	0·28370	0·30104	0·29857
0·912234	0·45436	0·44849	0·44433
0·999554	0·65425	0·55721	0·55181

3.11. Perturbation theory

We now turn to applications in quantum mechanics, and in this section we shall look at some results in perturbation theory which can be derived from complementary principles.

Consider a physical system described by the Schrödinger equation

$$(h - \epsilon_0)\psi_0 = 0, \tag{3.11.1}$$

where h is the Hamiltonian operator. The system is assumed to be in its ground state ψ_0 with energy ϵ_0. A perturbation V is applied to this and the first-order correction φ to ψ_0, by Rayleigh–Schrödinger theory [39], satisfies

$$(h - \epsilon_0)\varphi = (E_1 - V)\psi_0 \quad \text{in all space,} \tag{3.11.2}$$

with

$$\varphi = 0 \quad \text{at } \infty. \tag{3.11.3}$$

Here E_1 is the first-order correction to the unperturbed ground-state energy ϵ_0,

$$E_1 = \langle \psi_0, V\psi_0 \rangle, \tag{3.11.4}$$

and the second-order correction to ϵ_0 is given by

$$E_2 = \langle \varphi, (V - E_1)\psi_0 \rangle, \tag{3.11.5}$$

where ψ_0 is normalized and we are assuming for simplicity that all functions here are real.

Since $(h - \epsilon_0)$ is positive for all functions $\varphi \in D_h$, we can rewrite the boundary-value problem (3.11.2), (3.11.3) in the notation of section 3.2 by taking

$$T^*T = h - \epsilon_0, \tag{3.11.6}$$

$$Q = 0, \quad f = (E_1 - V)\psi_0, \quad \varphi_B = 0. \tag{3.11.7}$$

Then, by (3.2.49), we have

$$I(\varphi) \leqslant J(\Phi) \quad \Phi = 0 \text{ at } \infty, \tag{3.11.8}$$

where from (3.2.45)

$$J(\Phi) = \tfrac{1}{2}\langle \Phi, (h - \epsilon_0)\Phi \rangle - \langle (E_1 - V)\psi_0, \Phi \rangle, \tag{3.11.9}$$

and

$$I(\varphi) = J(\varphi) = \tfrac{1}{2}\langle (V - E_1)\psi_0, \varphi \rangle$$

$$= \tfrac{1}{2}E_2. \tag{3.11.10}$$

Using (3.11.10) in (3.11.8) we therefore obtain

$$E_2 \leqslant 2J(\Phi). \tag{3.11.11}$$

This result is known as the Hylleraas upper bound [40] for E_2, and from this derivation [16, 70] it is clear that its validity is independent of the way in which the decomposition is carried out.

To obtain a lower bound for E_2 which is complementary to the Hylleraas upper bound we write (3.11.2) in the form

$$\{(h - \epsilon_1) + (\epsilon_1 - \epsilon_0)\}\varphi = (E_1 - V)\psi_0, \tag{3.11.12}$$

where ϵ_1 ($\epsilon_1 > \epsilon_0$) is the unperturbed first excited-state energy. Now the right-hand side of (3.11.12) is orthogonal to ψ_0, by virtue of the value of E_1 in (3.11.4), and the solution φ of (3.11.12) is not unique, for

φ plus a constant multiple of ψ_0 is also a solution. Thus we may consider (3.11.12) as an equation on the domain D_0 of functions orthogonal to ψ_0, where it will have a unique solution φ. The operator $(h-\epsilon_1)$ is positive on D_0, and so we can set

$$T^*T = h-\epsilon_1, \qquad (3.11.13)$$

$$Q = \epsilon_1-\epsilon_0 > 0, \qquad f = (E_1-V)\psi_0, \qquad \varphi_B = 0. \quad (3.11.14)$$

From (3.2.50) we therefore find

$$G(T\Psi) \leqslant I(\varphi), \qquad (3.11.15)$$

where by (3.2.46)

$$G(T\Psi) = -\tfrac{1}{2}\langle \Psi, (h-\epsilon_1)\Psi \rangle - \frac{1}{2(\epsilon_1-\epsilon_0)} \times$$

$$\times \langle (E_1-V)\psi_0-(h-\epsilon_1)\Psi, (E_1-V)\psi_0-(h-\epsilon_1)\Psi \rangle. \quad (3.11.16)$$

Using (3.11.10) in (3.11.15) we obtain the lower bound

$$2G(T\Psi) \leqslant E_2. \qquad (3.11.17)$$

This result is due to Robinson [70]. In practice, if ϵ_1 is not known precisely, it can always be replaced by any quantity γ which satisfies

$$\epsilon_0 < \gamma < \epsilon_1. \qquad (3.11.18)$$

It is evident that any ψ_0-component in φ or in the trial functions Φ and Ψ would not contribute to any term in the upper and lower bounds, and so the restriction that $\Phi, \Psi \in D_0$ can be relaxed.

Other approaches to the lower-bound problem for E_2 have also been suggested [16, 64], but they are subject to rather stringent conditions and we shall not pursue them here.

The complementary bounds $2J(\Phi)$ and $2G(T\Psi)$ can be rewritten by using the Ritz procedure described in section 3.3. The resulting formulae are simplest if we expand

$$\Phi = \sum_{n=1}^{m} a_n \psi_n, \qquad \Psi = \sum_{n=1}^{m} b_n \psi_n \qquad (3.11.19)$$

in terms of the normalized eigenfunctions $\{\psi_n\}$ of h. Optimization with respect to the coefficients a_n and b_n yields

$$-\sum_{n=1}^{m} \frac{V_{n0}^2}{\epsilon_n-\epsilon_0} \geqslant E_2 \geqslant -\frac{1}{\epsilon_1-\epsilon_0} \times$$

$$\times \left\{ \langle (V-E_1)\psi_0, (V-E_1)\psi_0 \rangle - \sum_{n=1}^{m} \left(\frac{\epsilon_n-\epsilon_1}{\epsilon_n-\epsilon_0} \right) V_{n0}^2 \right\}, \quad (3.11.20)$$

where

$$V_{n0} = \langle \psi_n, V\psi_0 \rangle. \qquad (3.11.21)$$

More general formulae can be obtained with other basis sets. The left-hand member of (3.11.20) is well known, and taking $m = 1$ on the right gives a result due to Dalgarno [29].

Example. As a simple example we consider the Stark effect for hydrogen, where

$$\psi_0 = \pi^{-\frac{1}{2}}e^{-r}, \qquad V = -z, \qquad \epsilon_0 = -\tfrac{1}{2}, \qquad E_1 = 0. \qquad (3.11.22)$$

Taking trial functions of the form

$$\Phi = A\pi^{-\frac{1}{2}}ze^{-\alpha r}, \qquad \Psi = B\pi^{-\frac{1}{2}}ze^{-\beta r}, \qquad (3.11.23)$$

Robinson [70] obtained the bounds

$$-2{\cdot}271 \leqslant E_2 \leqslant -2{\cdot}238, \qquad (3.11.24)$$

with the optimum parameter values

$$B = 1{\cdot}455, \qquad \beta = 0{\cdot}662; \qquad A = 1{\cdot}310, \qquad \alpha = 0{\cdot}799. \qquad (3.11.25)$$

The exact result is

$$E_2 = -2{\cdot}25, \qquad \varphi = \pi^{-\frac{1}{2}}(1 + \tfrac{1}{2}r)ze^{-r}. \qquad (3.11.26)$$

3.12. Potential scattering

Our second application of complementary variational principles in quantum mechanics deals with the scattering of particles by a short-range central force. For background information on variational principles in the quantum theory of scattering we refer to chapter five of the book by Moiseiwitsch [54].

We shall consider a zero-energy potential scattering process, and for simplicity the orbital angular momentum will be taken to be zero. Then the s-wave $\varphi(r)$ can be specified in two equivalent ways. We can either regard $\varphi(r)$ as the solution of the differential equation

$$\left\{-\frac{d^2}{dr^2} + p(r)\right\}\varphi(r) = 0 \qquad 0 \leqslant r < \infty, \qquad (3.12.1)$$

subject to the conditions

$$\varphi(0) = 0, \qquad \varphi(r) \sim A - r \qquad \text{as } r \to \infty, \qquad (3.12.2)$$

or alternatively think of it as the solution of the integral equation

$$\varphi(r) = -r - \int_0^\infty \min(r, r')p(r')\varphi(r')\, dr'. \qquad (3.12.3)$$

Here

$$p(r) = \frac{2m}{\hbar^2}\, V(r), \qquad (3.12.4)$$

where $V(r)$ is a short-range potential and m is the mass of the scattered particle. The scattering length A in (3.12.2), which determines the cross-section at vanishing energy [cf. 54], is given by the relation

$$A = -\int\limits_0^\infty rp(r)\varphi(r)\,\mathrm{d}r. \tag{3.12.5}$$

Variational bounds for the scattering length are known from the work of Schwinger [48], who used the integral equation (3.12.3), and Spruch and Rosenberg [76], who used the differential equation (3.12.1). As we shall see, these results, together with their complementary companions obtained recently in [8, 11], follow directly from the general theory of section 3.2.

Differential equation approach

We first consider the scattering problem from the point of view of the differential equation (3.12.1) subject to the boundary conditions (3.12.2). It is convenient to treat the cases $p > 0$ and $p < 0$ separately.

Case 1: $p > 0$.

When p is positive we can take

$$T = \frac{\mathrm{d}}{\mathrm{d}r}, \qquad T^* = -\frac{\mathrm{d}}{\mathrm{d}r}, \qquad \sigma_T = 1, \tag{3.12.6}$$

$$Q = p, \qquad f = 0, \tag{3.12.7}$$

$$\varphi_B = 0 \quad \text{at } r = 0, \qquad \varphi_B \sim A - r \quad \text{as } r \to \infty. \tag{3.12.8}$$

Then the results of section 3.2 apply to (3.12.1), (3.12.2), provided condition (3.2.19) is satisfied. The optimum choice for this condition is

$$\left[(\Phi - \varphi_B)\frac{\mathrm{d}}{\mathrm{d}r}(\Phi - \varphi)\right]_0^\infty = 0. \tag{3.12.9}$$

We shall satisfy (3.12.9) by taking the trial function Φ such that

$$\Phi(0) = 0, \qquad \Phi \sim a - r \quad \text{as } r \to \infty, \tag{3.12.10}$$

where a is a constant. The basic functionals in (3.2.45) and (3.2.46) then become

$$J(\Phi) = \frac{1}{2}\left\langle \Phi, \left(-\frac{\mathrm{d}^2}{\mathrm{d}r^2} + p\right)\Phi \right\rangle + \tfrac{1}{2}(R - 2A + a)_{R \to \infty}, \tag{3.12.11}$$

$$I(\varphi) = \tfrac{1}{2}(R - A)_{R \to \infty} \tag{3.12.12}$$

$$G(T\Psi) = \frac{1}{2}\left\langle \Psi, \frac{\mathrm{d}^2}{\mathrm{d}r^2}\Psi \right\rangle - \frac{1}{2}\left\langle \frac{\mathrm{d}^2\Psi}{\mathrm{d}r^2}, p^{-1}\frac{\mathrm{d}^2\Psi}{\mathrm{d}r^2} \right\rangle - $$
$$- \left[\frac{\mathrm{d}\Psi}{\mathrm{d}r}(\tfrac{1}{2}\Psi - \varphi_B)\right]_0^\infty. \tag{3.12.13}$$

The boundary term involving R can be subtracted from each functional, and to get a useful bound from G we make the trial function Ψ satisfy boundary conditions of the form

$$\Psi(0) = 0, \qquad \Psi \sim b - r \quad \text{as} \quad r \to \infty, \tag{3.12.14}$$

otherwise the G bound recedes to minus infinity. Then from $G \leqslant I \leqslant J$ we obtain upper and lower bounds for the scattering length A, namely

$$A_-(\Psi) \leqslant A \leqslant A_+(\Phi), \tag{3.12.15}$$

where

$$A_+(\Phi) = a + \int_0^\infty \Phi\left(-\frac{d^2}{dr^2} + p\right)\Phi \, dr, \tag{3.12.16}$$

and

$$A_-(\Psi) = b + \int_0^\infty \frac{d^2\Psi}{dr^2}\left(1 - p^{-1}\frac{d^2}{dr^2}\right)\Psi \, dr. \tag{3.12.17}$$

The upper bound (3.12.16) is the one due to Spruch and Rosenberg [76], while the lower bound (3.12.17) was derived in [8].

Case 2: $p < 0$.

When p is negative we cannot set $Q = p$ as in case 1, because Q is to be positive. We retain the identification

$$-\frac{d^2}{dr^2} + p = T^*T + Q, \tag{3.12.18}$$

but this time we take

$$Q = (\lambda_0 - 1)(-p), \tag{3.12.19}$$

$$T^*T = -\frac{d^2}{dr^2} + \lambda_0 p, \tag{3.12.20}$$

where λ_0 is the lowest eigenvalue of the equation

$$(-p)^{-1}\left(-\frac{d^2}{dr^2}\right)\theta = \lambda_0\theta. \tag{3.12.21}$$

It is clear that Q is positive-definite provided $\lambda_0 > 1$. From (3.12.20) it follows that

$$T = \frac{d}{dr} + \tau(r), \qquad T^* = -\frac{d}{dr} + \tau(r), \tag{3.12.22}$$

where $\tau(r)$ is a short-range function of r which depends on $p(r)$. It is not necessary to find $\tau(r)$, since to evaluate the boundary terms in (3.2.45)–(3.2.48) we merely need to know the nature of T when r is large.

We now apply the theory of section 3.2 with

$$f = 0, \qquad \sigma_T = 1, \tag{3.12.23}$$

taking trial functions Φ and Ψ which satisfy the boundary conditions (3.12.10) and (3.12.14). The resulting bounds for A are readily found to be

$$A'_-(\Psi) \leqslant A \leqslant A'_+(\Phi), \tag{3.12.24}$$

where

$$A'_+(\Phi) = a + \int_0^\infty \Phi\left(-\frac{d^2}{dr^2} + p\right)\Phi \, dr, \tag{3.12.25}$$

and $A'_-(\Psi) = A'_+(\Psi) + (\lambda_0 - 1)^{-1}\int_0^\infty p^{-1}\left\{\left(-\frac{d^2}{dr^2} + p\right)\Psi\right\}^2 dr \quad (\lambda_0 > 1).$

$$\tag{3.12.26}$$

The upper bound (3.12.25) is that due to Spruch and Rosenberg [76], being identical to the expression in (3.12.16), while the lower bound (3.12.26) was derived in [8].

Integral equation approach

We next turn to the integral equation approach specified by equations (3.12.3) and (3.12.5). It is more convenient to rewrite (3.12.3) in the form

$$(K + p)\varphi = -rp, \tag{3.12.27}$$

where K is the symmetric, positive-definite integral operator defined by

$$K\psi(r) = \int_0^\infty p(r)\min\{r, r'\}p(r')\psi(r') \, dr'. \tag{3.12.28}$$

Equation (3.12.27) can be decomposed in various ways [8], with T and T^* equal to certain integral operators so that

$$\sigma_T = 0, \tag{3.12.29}$$

and no boundary terms appear in the basic functionals I, J, and G. This means that T and T^* only occur in the product T^*T, and individual representations of them are not required. Because of (3.12.29), the condition (3.2.19) is automatically satisfied.

Case 1: $p > 0$.

For positive potentials the straightforward choice is

$$T^*T = K, \qquad \sigma_T = 0,$$

$$Q = p, \qquad f = -rp. \tag{3.12.30}$$

Using expression (3.12.5) for A, and the functionals (3.2.45), (3.2.46), and (3.2.48), we find that $G \leqslant I \leqslant J$ leads to

$$B_-(\Phi) \leqslant A \leqslant B_+(\Psi), \tag{3.12.31}$$

where
$$B_+(\Psi) = \int_0^\infty \{\Psi K \Psi + p^{-1}(rp + K\Psi)^2\} \, dr, \tag{3.12.32}$$

and
$$B_-(\Phi) = -\int_0^\infty \{2rp\Phi + \Phi(K+p)\Phi\} \, dr. \tag{3.12.33}$$

The lower bound (3.12.33) is due to Schwinger [48], while the upper bound (3.12.32) was derived in [11].

Case 2: $p < 0$.

For negative potentials a suitable choice is

$$T^*T = -(p\lambda_0^{-1} + K), \qquad \sigma_T = 0,$$

$$Q = (1 - \lambda_0^{-1})(-p), \qquad f = rp, \tag{3.12.34}$$

where we now think of λ_0 as the smallest eigenvalue of

$$\lambda_0(-p)^{-1}K\theta = \theta. \tag{3.12.35}$$

Thus Q is positive-definite provided again that $\lambda_0 > 1$. Using (3.12.5) and (3.12.34) in the basic functionals (3.2.45), (3.2.46), and (3.2.48), we find that
$$B'_-(\Psi) \leqslant A \leqslant B'_+(\Phi), \tag{3.12.36}$$

where
$$B'_+(\Phi) = -\int_0^\infty \{2rp\Phi + \Phi(K+p)\Phi\} \, dr, \tag{3.12.37}$$

and $B'_-(\Psi) = \int_0^\infty \Psi(p\lambda_0^{-1} + K)\Psi \, dr +$

$$+ \frac{\lambda_0}{\lambda_0 - 1} \int_0^\infty p^{-1}\{rp + (p\lambda_0^{-1} + K)\Psi\}^2 \, dr \quad (\lambda_0 > 1). \tag{3.12.38}$$

The upper bound (3.12.37) is due to Schwinger [48], being identical to the expression (3.12.33), while the lower bound (3.12.38) is contained in a result of [11].

We have described here the main decompositions based on the differential and integral equation approaches. For an additional decomposition we refer to the results given in [8].

Examples. Screened Coulomb potentials.

To illustrate this theory we shall consider the calculation of bounds on scattering lengths for both positive and negative screened Coulomb potentials given by

$$V(r) = \pm \frac{e^{-\beta r}}{r},$$ (3.12.39)

β being some positive parameter. The scattered particle will be taken to have mass $m = 1$ atomic unit.

Following the work in [8] we take the trial functions

$$a(1 - e^{-\alpha r}) - r, \quad \text{differential equation approach,} \quad (3.12.40)$$

$$a(1 - e^{-r}) - r(1 - e^{-r}), \quad \text{integral equation approach,} \quad (3.12.41)$$

where a and α are variational parameters. These functions have the correct behaviour at zero and infinity. The trial function (3.12.41), containing a linear variational parameter, is one of the simplest functions that can be employed. It proves to be too inflexible in the differential equation approach as it leads to divergent lower bounds for $\beta \geqslant 2$, but this shortcoming is readily avoided by using the trial function (3.12.40). Calculations have been performed [3, 8] for a range of values of β and the results (in atomic units) are shown in Tables 3.6 and 3.7. For some bounds, A'_- and B'_-, the eigenvalue λ_0 was required. This was calculated by an iteration method, giving

$$\lambda_0 = (0.8397)\beta.$$ (3.12.42)

The condition $\lambda_0 > 1$, which must be satisfied in A'_- and B'_-, therefore places a lower limit on possible values of β in these cases.

Judging by the results in Tables 3.6 and 3.7 we see that in the positive potential case the B-bounds are better than the A-bounds, while in the negative potential case the A'_+ and B'_- bounds give the best results. In general, because of smoothing effects, we expect bounds involving integral operators to be better than bounds involving differential operators, and so for a given trial function the B-bounds should be better than the A-bounds.

TABLE 3.6. *Upper and lower bounds on scattering lengths for* $V = e^{-\beta r}/r$

β	A_-	A_+	B_-	B_+
1	1·0443	1·0595	1·0522	1·0587
2	0·33902	0·34058	0·34044	0·34053
3	0·16817	0·16855	0·16854	0·16854
5	0·066920	0·066980	0·066974	0·066974
10	0·018201	0·018206	0·018204	0·018204

TABLE 3.7. *Upper and lower bounds on scattering lengths for* $V = -e^{-\beta r}/r$

β	A_-	A'_+	B'_-	B'_+
2	−1·2359	−1·1030	−1·1121	−1·0967
3	−0·36850	−0·34321	−0·34321	−0·34232
5	−0·10501	−0·10082	−0·10114	−0·10072
10	−0·022704	−0·022260	−0·022260	−0·022255

Finally we note that some of the results obtained here can be extended to provide upper and lower bounds on phase shifts at non-zero energies [37].

SUMMARY

This chapter has been concerned with applications of the theory of Chapter 2 to a class of linear boundary-value problems. The associated complementary variational principles were illustrated by various examples taken from mathematical physics. These examples were in (1) eigenvalue theory, giving the Ritz and Temple bounds, (2) potential theory, where upper and lower bounds were obtained for the capacity of a surface, (3) electrostatics, leading to the Dirichlet and Thomson bounds, (4) magnetostatics, giving bounds for the self-interaction energy of a steady current, (5) diffusion, providing bounds for the absorption probability, the extrapolated end-point, and the translational diffusion constant, (6) quantum perturbation theory, leading to the Hylleraas upper bound for the second-order energy and its complementary companion, and (7) quantum scattering theory, giving the Spruch–Rosenberg and Schwinger bounds for scattering lengths together with their complementary bounds.

4

NONLINEAR APPLICATIONS

4.1. A class of nonlinear problems

THE general variational theory of Chapter 2 will now be applied to a certain class of nonlinear boundary-value problems described by the equations

$$T^*T\varphi = F(\varphi) \quad \text{in } V, \tag{4.1.1}$$

$$\sigma_T(\varphi - \varphi_B) = 0 \quad \text{on } \partial V. \tag{4.1.2}$$

Here $T: H_\varphi \to H_u$ and its adjoint $T^*: H_u \to H_\varphi$ are linear operators of the kind discussed in Chapter 2 such that

$$(u, T\varphi) = \langle T^*u, \varphi \rangle + [u, \sigma_T \varphi], \tag{4.1.3}$$

and $F: H_\varphi \to H_\varphi$ is a given operator which may be nonlinear. The linear problems treated in Chapter 3 correspond to the special choice $F(\varphi) = -Q\varphi + f$.

Several problems of physical interest are described by equations of the form (4.1.1), (4.1.2), and so it is convenient for later applications to develop the corresponding variational theory at this point. We shall assume that the problem in (4.1.1), (4.1.2) has a solution and that it is unique.

4.2. Associated variational theory

We wish to apply the general canonical theory of Chapter 2 to the equations (4.1.1), (4.1.2), and to do this we must write (4.1.1) as a pair of canonical equations. A suitable form is given by

$$T\varphi = u = \delta W / \delta u, \tag{4.2.1}$$

$$T^*u = F(\varphi) = \delta W / \delta \varphi, \tag{4.2.2}$$

and, by direct integration as in (2.5.40), we see that a suitable W is

$$W(u, \varphi) = \tfrac{1}{2}(u, u) + \int_0^1 \langle \varphi - \varphi_0, F\{\varphi_0 + t(\varphi - \varphi_0)\}\rangle \, dt + W(0, \varphi_0), \tag{4.2.3}$$

where φ_0 is an arbitrary vector in H_φ. Having found $W(u, \varphi)$ we can now obtain the potential $I(u, \varphi)$ of the variational theory from (2.5.14) and (2.5.15). We find that

$$I(U, \Phi) = (U, T\Phi) - \tfrac{1}{2}(U,U) - \int_0^1 \langle \Phi - \varphi_0, F\{\varphi_0 + t(\Phi - \varphi_0)\} \rangle \, dt -$$

$$- W(0, \varphi_0) - [U, \sigma_T(\Phi - \varphi_B)], \quad (4.2.4)$$

$$= \langle T^*U, \Phi \rangle - \tfrac{1}{2}(U, U) - \int_0^1 \langle \Phi - \varphi_0, F\{\varphi_0 + t(\Phi - \varphi_0)\} \rangle \, dt -$$

$$- W(0, \varphi_0) + [U, \sigma_T \varphi_B]. \quad (4.2.5)$$

At this stage it is clear that for a nonlinear $F(\varphi)$ the corresponding potential $I(u, \varphi)$ contains terms other than linear and quadratic terms. This in turn means that third and higher order variations may not vanish, in contrast to the situation for linear $F(\varphi)$ treated in Chapter 3.

If (u, φ) denotes the exact solution of (4.2.1) and (4.2.2) subject to (4.1.2), it follows from (4.2.4) or (4.2.5) that

$$I(u, \varphi) = \tfrac{1}{2}\langle \varphi, F(\varphi) \rangle - \int_0^1 \langle \varphi - \varphi_0, F\{\varphi_0 + t(\varphi - \varphi_0)\} \rangle \, dt - W(0, \varphi_0) +$$

$$+ \tfrac{1}{2}[T\varphi, \sigma_T \varphi_B] \quad (4.2.6)$$

is the solution of the variational problem. For $F(\varphi) = -Q\varphi + f$, this reduces to (3.2.6) as it should, on taking $\varphi_0 = 0$ and $W(0, 0) = 0$.

From theorem 2.5.3 we know that under certain circumstances it is possible to obtain upper and lower bounds for the value of $I(u, \varphi)$. To discuss this we derive the basic functionals G and J.

First we derive J, which by (2.5.37) is defined by

$$J(\Phi) = I(Y(\Phi), \Phi), \quad (4.2.7)$$

where it is assumed that the canonical equation $T\Phi = \delta W/\delta U$ has the solution
$$U = Y(\Phi). \quad (4.2.8)$$

From (4.2.1) this means that

$$Y(\Phi) = T\Phi, \quad (4.2.9)$$

and so
$$J(\Phi) = I(T\Phi, \Phi). \quad (4.2.10)$$

By (4.2.4) we therefore find that

$$J(\Phi) = \tfrac{1}{2}(T\Phi, T\Phi) - \int_0^1 \langle \Phi - \varphi_0, F\{\varphi_0 + t(\Phi - \varphi_0)\} \rangle \, dt - W(0, \varphi_0) -$$

$$- [T\Phi, \sigma_T(\Phi - \varphi_B)], \quad (4.2.11)$$

$$= \tfrac{1}{2}\langle \Phi, T^*T\Phi \rangle - \int_0^1 \langle \Phi - \varphi_0, F\{\varphi_0 + t(\Phi - \varphi_0)\} \rangle \, dt - W(0, \varphi_0) -$$

$$- [T\Phi, \sigma_T(\tfrac{1}{2}\Phi - \varphi_B)]. \quad (4.2.12)$$

If we expand $J(\Phi)$ about the exact function φ we obtain

$$J(\Phi) = I(u, \varphi) + \delta^2 J(\Phi) + \delta^3 J(\Phi) + \ldots, \qquad (4.2.13)$$

where from (2.5.38)

$$\delta^2 J = \frac{1}{2}\left(Y - u, \frac{\delta^2 W}{\delta u^2}(Y - u)\right) - \frac{1}{2}\left\langle \Phi - \varphi, \frac{\delta^2 W}{\delta \varphi^2}(\Phi - \varphi)\right\rangle -$$
$$- [Y - u, \sigma_T(\Phi - \varphi_B)]. \quad (4.2.14)$$

The second derivatives of W are

$$\frac{\delta^2 W}{\delta u^2} = 1, \qquad \frac{\delta^2 W}{\delta \varphi^2} = \frac{\mathrm{d}F}{\mathrm{d}\varphi} \qquad (4.2.15)$$

from (4.2.3). Setting these in (4.2.14) and using $Y = T\Phi$, $u = T\varphi$, we obtain

$$\delta^2 J = \frac{1}{2}(T\Phi - T\varphi, T\Phi - T\varphi) - \frac{1}{2}\left\langle \Phi - \varphi, \frac{\mathrm{d}F}{\mathrm{d}\varphi}(\Phi - \varphi)\right\rangle -$$
$$- [T(\Phi - \varphi), \sigma_T(\Phi - \varphi_B)]. \quad (4.2.16)$$

We note that this expression also follows directly from (4.2.11). For Φ close to φ, so that third and higher variations are negligible, we see from (4.2.13) that

$$I(u, \varphi) \leqslant J(\Phi), \qquad (4.2.17)$$

if

$$\delta^2 J \geqslant 0. \qquad (4.2.18)$$

This condition (4.2.18) is certainly satisfied if

$$-\frac{\mathrm{d}F}{\mathrm{d}\varphi} \text{ is positive,} \qquad (4.2.19)$$

and if

$$[T(\Phi - \varphi), \sigma_T(\Phi - \varphi_B)] \leqslant 0. \qquad (4.2.20)$$

These sufficient conditions will be used later in applications. Condition (4.2.19) is sufficient to ensure the uniqueness of φ in many cases.

Now we turn to the complementary principle involving the functional $G(U)$. By (2.5.37) G is defined as

$$G(U) = I(U, \Theta(U)), \qquad (4.2.21)$$

where it is assumed that the second canonical equation $T^*U = \delta W/\delta\Phi$ has the solution

$$\Phi = \Theta(U). \qquad (4.2.22)$$

From (4.2.2) this means that

$$\Theta(U) = F^{-1}(T^*U), \qquad (4.2.23)$$

where we suppose that the inverse F^{-1} of F exists. Using (4.2.22) we therefore have

$$G(U) = I(U, F^{-1}(T^*U)) \qquad (4.2.24)$$

which, by (4.2.5), gives

$$G(U) = \langle T^*U, F^{-1}(T^*U)\rangle - \tfrac{1}{2}(U, U) -$$
$$- \int_0^1 \langle F^{-1}(T^*U) - \varphi_0, F\{\varphi_0 + t(F^{-1}(T^*U) - \varphi_0)\}\rangle \, dt -$$
$$- W(0, \varphi_0) + [U, \sigma_T \varphi_B]. \qquad (4.2.25)$$

If we expand $G(U)$ about the exact function u we obtain

$$G(U) = I(u, \varphi) + \delta^2 G(U) + \delta^3 G(U) + \dots, \qquad (4.2.26)$$

where from (2.5.39)

$$\delta^2 G = -\frac{1}{2}\left\{\left(U - u, \frac{\delta^2 W}{\delta u^2}(U - u)\right) - \left\langle \Theta - \varphi, \frac{\delta^2 W}{\delta \varphi^2}(\Theta - \varphi)\right\rangle\right\}$$
$$= -\frac{1}{2}\left\{(U - u, U - u) - \left\langle \Theta - \varphi, \frac{dF}{d\varphi}(\Theta - \varphi)\right\rangle\right\}, \qquad (4.2.27)$$

by (4.2.15). For U close to u, so that third and higher variations are negligible, we see from (4.2.26) that

$$G(U) \leqslant I(u, \varphi), \qquad (4.2.28)$$

if

$$\delta^2 G \leqslant 0. \qquad (4.2.29)$$

From (4.2.2) we see that this condition is certainly satisfied if

$$-\frac{dF}{d\varphi} \text{ is positive.} \qquad (4.2.30)$$

The result in (4.2.28) is the complementary principle of (4.2.17).

Since the exact function u is related to φ by $u = T\varphi$ as in (4.2.1), it is convenient to choose the trial function U to have the form

$$U = T\Psi, \qquad (4.2.31)$$

where $\Psi \in D_T$ is an approximation to φ. We then find from (4.2.25) that

$$G(T\Psi) = \langle T^*T\Psi, F^{-1}(T^*T\Psi)\rangle - \tfrac{1}{2}(T\Psi, T\Psi)$$
$$- \int_0^1 \langle F^{-1}(T^*T\Psi) - \varphi_0, F\{\varphi_0 + t(F^{-1}(T^*T\Psi) - \varphi_0)\}\rangle \, dt$$
$$- W(0, \varphi_0) + [T\Psi, \sigma_T \varphi_B], \qquad (4.2.32)$$
$$= -\tfrac{1}{2}\langle \Psi, T^*T\Psi\rangle + \langle T^*T\Psi, F^{-1}(T^*T\Psi)\rangle$$
$$- \int_0^1 \langle F^{-1}(T^*T\Psi) - \varphi_0, F\{\varphi_0 + t(F^{-1}(T^*T\Psi) - \varphi_0)\}\rangle \, dt$$
$$- W(0, \varphi_0) - [T\Psi, \sigma_T(\tfrac{1}{2}\Psi - \varphi_B)]. \qquad (4.2.33)$$

These are the basic results for the class of nonlinear problems in (4.1.1) and (4.1.2), and for reference purposes it is convenient to summarize them here.

SUMMARY

1. *Nonlinear problem*

$$T^*T\varphi = F(\varphi) \quad \text{in } V, \tag{4.2.34}$$

$$\sigma_T(\varphi - \varphi_B) = 0 \quad \text{on } \partial V, \tag{4.2.35}$$

where F is an operator, nonlinear in general, T^* is the adjoint of the linear operator T defined by (4.1.3),

$$(u, T\varphi) = \langle T^*u, \varphi\rangle + [u, \sigma_T\varphi], \tag{4.2.36}$$

and φ_B prescribes the value of the exact solution φ on the boundary ∂V of some region V.

2. *Associated variational problem*

The basic functionals are

$$J(\Phi) = \tfrac{1}{2}(T\Phi, T\Phi) - \int_0^1 \langle \Phi - \varphi_0, F\{\varphi_0 + t(\Phi - \varphi_0)\}\rangle \, dt - W(0, \varphi_0) -$$
$$- [T\Phi, \sigma_T(\Phi - \varphi_B)], \tag{4.2.37}$$

$$= \tfrac{1}{2}\langle \Phi, T^*T\Phi\rangle - \int_0^1 \langle \Phi - \varphi_0, F\{\varphi_0 + t(\Phi - \varphi_0)\}\rangle \, dt - W(0, \varphi_0)$$
$$- [T\Phi, \sigma_T(\tfrac{1}{2}\Phi - \varphi_B)]. \tag{4.2.38}$$

$$G(T\Psi) = -\tfrac{1}{2}(T\Psi, T\Psi) + \langle T^*T\Psi, F^{-1}(T^*T\Psi)\rangle -$$
$$- \int_0^1 \langle F^{-1}(T^*T\Psi) - \varphi_0, F\{\varphi_0 + t(F^{-1}(T^*T\Psi) - \varphi_0)\}\rangle \, dt -$$
$$- W(0, \varphi_0) + [T\Psi, \sigma_T\varphi_B], \tag{4.2.39}$$

$$= -\tfrac{1}{2}\langle \Psi, T^*T\Psi\rangle + \langle T^*T\Psi, F^{-1}(T^*T\Psi)\rangle -$$
$$- \int_0^1 \langle F^{-1}(T^*T\Psi) - \varphi_0, F\{\varphi_0 + t(F^{-1}(T^*T\Psi) - \varphi_0)\}\rangle \, dt -$$
$$- W(0, \varphi_0) - [T\Psi, \sigma_T(\tfrac{1}{2}\Psi - \varphi_B)]. \tag{4.2.40}$$

$$I(\varphi) \equiv I(u, \varphi) = \tfrac{1}{2}\langle \varphi, F(\varphi)\rangle - \int_0^1 \langle \varphi - \varphi_0, F\{\varphi_0 + t(\varphi - \varphi_0)\}\rangle \, dt -$$
$$- W(0, \varphi_0) + \tfrac{1}{2}[T\varphi, \sigma_T\varphi_B]. \tag{4.2.41}$$

3. *Extremum principles:*

Minimum principle

$$I(\varphi) \leqslant J(\Phi) \tag{4.2.42}$$

if $$-\frac{\mathrm{d}F}{\mathrm{d}\varphi} \text{ is positive} \tag{4.2.43}$$

$$[T(\Phi-\varphi),\ \sigma_T(\Phi-\varphi_B)] \leqslant 0, \tag{4.2.44}$$

and $\delta^n J$ is negligible, or vanishes, for $n = 3, 4, \ldots$ (4.2.45)

Maximum principle

$$G(T\Psi) \leqslant I(\varphi) \tag{4.2.46}$$

if $$-\frac{\mathrm{d}F}{\mathrm{d}\varphi} \text{ is positive,} \tag{4.2.47}$$

and $\delta^n G$ is negligible, or vanishes, for $n = 3, 4, \ldots$ (4.2.48)

A slightly more general form of these results has been given in [12].

4.3. Liouville equation

As our first application of the variational principles developed in the previous section we shall consider the problem of finding the solution φ of the Liouville equation

$$\nabla^2\varphi = ce^\varphi \quad \text{in } V, \tag{4.3.1}$$

with

$$\varphi = \varphi_B \quad \text{on } \partial V. \tag{4.3.2}$$

Here c is a positive constant, and φ_B is a real continuous function. Equation (4.3.1) arises for example in the theory of charged particles in equilibrium [cf. 50], and in hydrodynamics [21, 30]. To use the results of section 4.2 we choose

$$T = \mathrm{grad}, \qquad T^* = -\mathrm{div}, \qquad \sigma_T = \mathbf{n}, \tag{4.3.3}$$

$$F(\varphi) = -ce^\varphi. \tag{4.3.4}$$

Since $\mathrm{d}F/\mathrm{d}\varphi = -ce^\varphi$, and c is positive, we see that the condition in (4.2.43) and (4.2.47) holds. Hence complementary variational principles can be obtained [17]. From (4.2.42) and (4.2.46) we find

$$G(T\Psi) \leqslant I(\varphi) \leqslant J(\Phi), \tag{4.3.5}$$

where, by (4.2.37), (4.2.39), and (4.2.41), with $\varphi_0 = 0$ and

$$W(0, 0) = -\langle c, 1 \rangle$$

to simplify the functionals,

$$J(\Phi) = \tfrac{1}{2}(\text{grad } \Phi, \text{grad } \Phi) + \langle c, \exp \Phi \rangle, \qquad \Phi = \varphi_B \text{ on } \partial V,$$
(4.3.6)

$$I(\varphi) = \tfrac{1}{2}(\text{grad } \varphi, \text{grad } \varphi) + \langle c, \exp \varphi \rangle,$$
(4.3.7)

$$G(T\Psi) = -\tfrac{1}{2}(\text{grad } \Psi, \text{grad } \Psi) - \langle \nabla^2\Psi, \ln(c^{-1}\nabla^2\Psi) \rangle +$$
$$+ \langle 1, \nabla^2\Psi \rangle + [\text{grad } \Psi, \mathbf{n}\varphi_B].$$
(4.3.8)

It is assumed here that Φ and Ψ are sufficiently close to the exact function φ.

Example.   Plasma theory.

To illustrate these results we take an example from the theory of charged particles in equilibrium [cf. 50]. In this theory the Liouville equation (4.3.1) arises when the Boltzmann number density of particles

$$n = n_0 \exp(-q\Omega/kT)$$
(4.3.9)

is substituted into the Poisson equation

$$\nabla^2\Omega = -4\pi nq$$
(4.3.10)

and we set

$$\Phi = -q\Omega/kT.$$
(4.3.11)

The charge per particle is q, the electrostatic potential is Ω, and kT is the Boltzmann energy. If we take the unit of length as $\sqrt{8}L_D$, where

$$L_D = \{kT/4\pi q^2 n_0\}^{\tfrac{1}{2}}$$
(4.3.12)

is the Debye shielding length, the constant c in (4.3.1) is given by

$$c = 8.$$
(4.3.13)

Let V be the volume per unit length between two long coaxial cylinders

$$V = \{a \leqslant r \leqslant b; 0 \leqslant \theta \leqslant 2\pi; z_1 \leqslant z \leqslant z_2\},$$
(4.3.14)

where (r, θ, z) are cylindrical polar coordinates. The boundary conditions are taken as

$$\varphi_B = \alpha \quad \text{on } r = a, \qquad \varphi_B = \beta \quad \text{on } r = b,$$
(4.3.15)

where α and β are constants. As trial functions we choose

$$\Phi = \xi + \eta r,$$
(4.3.16)

where, to give $\Phi = \varphi_B$ on ∂V, we take

$$\xi = \frac{b\alpha - a\beta}{b - a}, \qquad \eta = \frac{\beta - \alpha}{b - a},$$
(4.3.17)

and

$$\text{grad } \Psi = \left(\omega + \frac{\mu}{r}\right), \hat{\mathbf{r}}$$
(4.3.18)

where ω and μ are variational parameters determined from the stationary conditions $\partial G/\partial \omega = 0$ and $\partial G/\partial \mu = 0$. With these trial functions the functionals J and G are readily evaluated.

An interesting special case arises when

$$\alpha = -2\ln(a^2 - 1), \qquad \beta = -2\ln(b^2 - 1), \qquad (4.3.19)$$

because for this the exact solution φ is known, namely

$$\varphi = -2\ln(r^2 - 1). \qquad (4.3.20)$$

The corresponding exact functional $I(\varphi)$ is therefore known. For example, when $a = 2$, $b = 3$ we find [17] that

$$G = 11{\cdot}02\pi, \qquad I = 11{\cdot}18\pi, \qquad J = 11{\cdot}44\pi, \qquad (4.3.21)$$

showing that G and J are good bounds for I.

4.4. Poisson–Boltzmann equation

For our next application of the theory developed in section 4.2 we shall consider the problem of solving the Poisson–Boltzmann equation

$$\frac{\mathrm{d}^2\varphi}{\mathrm{d}x^2} = e^\varphi - e^{-\varphi} \qquad 0 \leqslant x \leqslant a, \qquad (4.4.1)$$

with boundary conditions

$$\varphi(0) = \varphi_1, \qquad \varphi(a) = \varphi_2 \qquad (4.4.2)$$

where φ_1 and φ_2 are given numbers. Such a problem arises for example in the theory of colloids [1, 33] ($a = \infty$, $\varphi_1 > 0$, $\varphi_2 = 0$) and in the theory of plasmas [49] (a finite, $\varphi_1 = 0$, $\varphi_2 > 0$). In these cases φ is the dimensionless quantity $q\Omega/kT$, where Ω is the electric potential, q is the proton charge, kT is the Boltzmann energy, and x is measured in units of the Debye length defined in equation (4.3.12).

To use the results of section 4.2 we choose

$$T = \frac{\mathrm{d}}{\mathrm{d}x}, \qquad T^* = -\frac{\mathrm{d}}{\mathrm{d}x}, \qquad \sigma_T = 1, \qquad (4.4.3)$$

$$F(\varphi) = -(e^\varphi - e^{-\varphi}), \qquad \varphi_B = \varphi_1 \text{ at } x = 0, \ \varphi_2 \text{ at } x = a. \quad (4.4.4)$$

With this $F(\varphi)$ we see that the condition in (4.2.43) and (4.2.47) holds, and so complementary variational principles can be found [2]. From (4.2.42) and (4.2.46) we get

$$G(T\Psi) \leqslant I(\varphi) \leqslant J(\Phi), \qquad (4.4.5)$$

where, by (4.2.37), (4.2.39), and (4.2.41), with $\varphi_0 = 0$ and

$$W(0, 0) = 0$$

to simplify the functionals,

$$J(\Phi) = \int_0^a \left\{ \frac{1}{2}\left(\frac{d\Phi}{dx}\right)^2 + 2 \cosh \Phi - 2 \right\} dx, \quad \Phi(0) = \varphi_1, \ \Phi(a) = \varphi_2, \quad (4.4.6)$$

$$I(\varphi) = \int_0^a \left\{ \frac{1}{2}\left(\frac{d\varphi}{dx}\right)^2 + 2 \cosh \varphi - 2 \right\} dx, \quad\quad\quad (4.4.7)$$

$$G(T\Psi) = \int_0^a \left\{ -\frac{1}{2}\left(\frac{d\Psi}{dx}\right)^2 - \frac{d^2\Psi}{dx^2} \sinh^{-1}\left(\frac{1}{2}\frac{d^2\Psi}{dx^2}\right) + \right.$$

$$\left. + 2 \cosh\left[\sinh^{-1}\left(\frac{1}{2}\frac{d^2\Psi}{dx^2}\right)\right] - 2 \right\} dx + \left[\frac{d\Psi}{dx} \varphi_B\right]_0^a. \quad (4.4.8)$$

It is assumed here that Φ and Ψ are sufficiently close to the exact function φ.

Example 1.   *Colloid problem.*

To illustrate these results we consider the boundary-value problem

$$\frac{d^2\varphi}{dx^2} = e^\varphi - e^{-\varphi} \quad\quad x > 0 \quad\quad\quad (4.4.9)$$

$$\varphi(0) = \varphi_1, \quad \lim_{a\to\infty} \varphi(a) = 0. \quad\quad\quad (4.4.10)$$

This problem arises in the Debye–Hückel theory of colloids [1]. When φ is much smaller than unity the solution is

$$\varphi \sim \varphi_1 e^{-x\sqrt{2}}, \quad \varphi_1 \ll 1. \quad\quad\quad (4.4.11)$$

This suggests that for $\varphi_1 \sim 1$ we choose the trial functions

$$\Phi = \varphi_1 e^{-\lambda x}, \quad \Psi = \varphi_1 e^{-\mu x}, \quad\quad\quad (4.4.12)$$

where the parameters λ and μ are determined by $\partial J/\partial \lambda = 0$, $\partial G/\partial \mu = 0$. Taking the case $\varphi_1 = 1$, we find that it is sufficient to have $a = 10$, and the results are [2]

$$\lambda = 1\cdot45, \quad \mu = 1\cdot48 \quad\quad\quad (4.4.13)$$

and

$$G = 0\cdot7213, \quad J = 0\cdot7222. \quad\quad\quad (4.4.14)$$

The numerical values in (4.4.14) provide upper and lower bounds for $I(\varphi)$ in (4.4.7), which for one-dimensional problems is a measure of the field energy. Thus, in terms of the field energy, the trial function

$$\Phi = e^{-1 \cdot 45x} \qquad (4.4.15)$$

is quite an accurate representation of the potential in this problem.

As φ_1 decreases from unity it is clear that the parameter λ in (4.4.12) decreases from 1·45 and tends to $\sqrt{2}$.

Example 2. Plasma problem.

As a second example of the Poisson–Boltzmann equation we consider

$$\frac{d^2\varphi}{dx^2} = e^{\varphi} - e^{-\varphi} \qquad 0 \leqslant x \leqslant a, \qquad (4.4.16)$$

subject to $\qquad\qquad \varphi(0) = 0, \qquad \varphi(a) = \varphi_2. \qquad (4.4.17)$

This problem arises in plasma theory [49]. The form of the exact solution near $x = a$ suggests that we take the trial functions

$$\Phi = \varphi_2 e^{-\alpha(a-x)} \{1 - e^{-\alpha x}\} \{1 - e^{-\alpha a}\}^{-1}, \qquad (4.4.18)$$

$$\Psi = \varphi_2 e^{-\beta(a-x)} \{1 - e^{-\beta x}\} \{1 - e^{-\beta a}\}^{-1}, \qquad (4.4.19)$$

where α and β are variational parameters determined by $\partial J/\partial \alpha = 0$, $\partial G/\partial \beta = 0$. Calculations have been performed [2] for $\varphi_2 = 1$ and $\varphi_2 = 2$, and $a = 1 \ (1 \cdot 0) \ 5$, and the results are given in Tables 4.1 and 4.2.

TABLE 4.1. *Parameters for $\varphi_2 = 1$*

a	α	J	β	G	$J - G$
1	0·93	0·8097	1.06	0·7785	0·0312
2	1.31	0·7278	1·42	0·7226	0·0052
3	1·39	0·7227	1·46	0·7212	0·0015
4	1·43	0·7222	1·48	0·7211	0·0011
5	1·44	0·7222	1·48	0·7213	0·0009

TABLE 4.2. *Parameters for $\varphi_2 = 2$*

a	α	J	β	G	$J - G$
1	1·05	3·4003	1·23	3·2148	0·1855
2	1·43	3·0993	1·61	3·0467	0·0526
3	1·51	3·0792	1·65	3·0470	0·0322
4	1·53	3·0771	1·66	3·0476	0·0295
5	1·54	3·0767	1·67	3·0479	0·0288

Since the accuracy of the trial functions is judged by the closeness of the bounds G and J [cf. 75], we see that the function (4.4.18) improves as a increases and φ_2 decreases, and provides a reasonable representation of the potential in this problem.

4.5. Thomas–Fermi equation

The Thomas–Fermi equation for an atomic system with a single nucleus of charge Z at the origin is [47] in atomic units

$$\nabla^2\Omega = 4\pi\lambda^{-\frac{3}{2}}(\Omega-\omega_0)^{\frac{3}{2}}-4\pi Z\delta(\mathbf{r}), \qquad (4.5.1)$$

where

$$\lambda = \tfrac{1}{2}(3\pi^2)^{\frac{2}{3}}, \qquad (4.5.2)$$

and the electrostatic potential $\Omega(\mathbf{r})$ satisfies the boundary condition

$$\Omega(\mathbf{r}) \to \omega_0 = \text{constant} \qquad \text{as } r \to \infty. \qquad (4.5.3)$$

If the atomic system is neutral, then

$$\omega_0 = 0. \qquad (4.5.4)$$

Setting

$$\Omega(\mathbf{r}) = \omega_0 + \frac{Z}{r}\,\varphi(\mathbf{r}), \qquad (4.5.5)$$

and

$$r = bx, \qquad (4.5.6)$$

where

$$b = \tfrac{1}{2}(\tfrac{3}{4}\pi)^{\frac{2}{3}}Z^{-\frac{1}{3}} = 0\cdot88534Z^{-\frac{1}{3}}, \qquad (4.5.7)$$

we find that the spherically symmetric form of (4.5.1), which describes the normal state of the system, becomes

$$\frac{\mathrm{d}^2\varphi}{\mathrm{d}x^2} = \frac{\varphi^{\frac{3}{2}}}{x^{\frac{1}{2}}} \qquad 0 \leqslant x < \infty. \qquad (4.5.8)$$

The boundary conditions are

$$\varphi(0) = 1; \qquad \varphi \to 0, \qquad x\varphi' \to 0 \quad \text{as } x \to \infty. \qquad (4.5.9)$$

These conditions follow from the consistency relation

$$\int n(\mathbf{r})\,\mathrm{d}\mathbf{r} = Z, \qquad (4.5.10)$$

which the electron number density

$$n(\mathbf{r}) = \lambda^{-\frac{3}{2}}(\Omega-\omega_0)^{\frac{3}{2}} = \left\{\frac{Z}{\lambda}\frac{\varphi(\mathbf{r})}{r}\right\}^{\frac{3}{2}} \qquad (4.5.11)$$

satisfies for neutral atoms. Since $n(\mathbf{r})$ is a nonnegative quantity, it is clear from (4.5.11) that the function $\varphi(\mathbf{r})$ is also nonnegative.

Complementary variational principles have been derived [17] for the Thomas–Fermi equation describing a system containing an arbitrary number of electrons and nuclei. If we restrict our attention here to a neutral atom in its ground state, the problem reduces to solving (4.5.8) subject to (4.5.9). This problem corresponds to

$$T = \frac{\mathrm{d}}{\mathrm{d}x}, \qquad T^* = -\frac{\mathrm{d}}{\mathrm{d}x}, \qquad \sigma_T = 1, \tag{4.5.12}$$

$$F(\varphi) = -\frac{\varphi^{\frac{3}{2}}}{x^{\frac{1}{2}}}, \qquad \varphi_B = 1 \text{ at } x = 0, \quad \varphi_B \to 0 \text{ as } x \to \infty. \tag{4.5.13}$$

Since φ is nonnegative we see from this $F(\varphi)$ that $\mathrm{d}F/\mathrm{d}\varphi \leqslant 0$, and hence complementary principles can be obtained. From (4.2.42) and (4.2.46) we have

$$G(T\Psi) \leqslant I(\varphi) \leqslant J(\Phi), \tag{4.5.14}$$

where, by (4.2.37), (4.2.39), and (4.2.41), with $\varphi_0 = 0$ and $W(0, 0) = 0$

$$J(\Phi) = \int_0^\infty \left\{ \frac{1}{2}\left(\frac{\mathrm{d}\Phi}{\mathrm{d}x}\right)^2 + \frac{2}{5}\frac{\Phi^{\frac{5}{2}}}{x^{\frac{1}{2}}} \right\} \mathrm{d}x$$

$$(\Phi(0) = 1; \Phi \to 0, x\Phi' \to 0 \quad \text{as } x \to \infty), \tag{4.5.15}$$

$$I(\varphi) = \int_0^\infty \left\{ \frac{1}{2}\left(\frac{\mathrm{d}\varphi}{\mathrm{d}x}\right)^2 + \frac{2}{5}\frac{\varphi^{\frac{5}{2}}}{x^{\frac{1}{2}}} \right\} \mathrm{d}x = -\frac{bE}{Z^2}, \tag{4.5.16}$$

E being the total electron energy,

$$G(T\Psi) = -\int_0^\infty \left\{ \frac{1}{2}\left(\frac{\mathrm{d}\Psi}{\mathrm{d}x}\right)^2 + \frac{3}{5} x^{\frac{1}{3}}\left(\frac{\mathrm{d}^2\Psi}{\mathrm{d}x^2}\right)^{\frac{5}{3}} \right\} \mathrm{d}x - \left[\frac{\mathrm{d}\Psi}{\mathrm{d}x}\right]_{x=0}. \tag{4.5.17}$$

Here the trial functions Φ and Ψ are intended to be close to the exact solution φ of (4.5.12) and (4.5.13).

The simplest kind of suitable trial function is given by

$$\Phi = \mathrm{e}^{-\alpha x}, \qquad \Psi = \mathrm{e}^{-\beta x}, \tag{4.5.18}$$

where α and β are variational parameters. With these functions the functionals G and J can be evaluated analytically to give

$$J(\Phi) = \frac{\alpha}{4} + \frac{2}{5}\left(\frac{2\pi}{5\alpha}\right)^{\frac{1}{2}}, \qquad G(T\Psi) = \tfrac{3}{4}\beta - \left(\frac{3}{5}\right)^{7/3} \Gamma\left(\frac{4}{3}\right)\beta^2. \tag{4.5.19}$$

The optimum values of these are

$$J = 0\cdot69732 \quad \text{at} \quad \alpha = 0\cdot930, \tag{4.5.20}$$

and

$$G = 0\cdot51863 \quad \text{at} \quad \beta = 1\cdot383. \tag{4.5.21}$$

For good trial functions the parameters α and β would be close and give rise to close bounds. Thus we see that these simple trial functions are not very satisfactory. For better results, more complicated functions must be considered.

One such function is

$$\Phi = (1+\gamma x^{\frac{1}{2}})e^{-\gamma x^{\frac{1}{2}}}, \qquad (4.5.22)$$

which has been proposed by Roberts [69], who determined γ by a method equivalent to optimizing the upper bound $J(\Phi)$. The lower bound $G(T\Psi)$ has also been studied [5] with a trial function Ψ of the same form as (4.5.22). The optimum values were found to be

$$J = 0\cdot6810 \quad \text{at} \quad \gamma = 1\cdot905, \qquad (4.5.23)$$

$$G = 0\cdot6699 \quad \text{at} \quad \gamma = 1\cdot750. \qquad (4.5.24)$$

These values represent a marked improvement on those in (4.5.20) and (4.5.21), and indicate that the trial function in (4.5.22) is a good one. This is confirmed by a comparison of (4.5.22) with the numerical solution [24, 44] of (4.5.8) and (4.5.9). The numerical solution of course also provides the exact value of $I(\varphi)$ sandwiched by G and J, and this is

$$I(\varphi) = -\frac{bE}{Z^2} = 0\cdot6806. \qquad (4.5.25)$$

4.6. A nonlinear integral equation

The nonlinear integral equation

$$\varphi K\varphi = 1, \qquad (4.6.1)$$

where

$$K\psi(x) = \int_0^{\pi/2} \frac{\sin(x-y)}{\pi(x-y)} \, \psi(y) \, dy, \qquad 0 \leqslant x \leqslant \frac{\pi}{2}, \qquad (4.6.2)$$

arises in communication theory [58, 72], the solution φ being the Fourier transform of the transmission signal. In (4.6.2) K is a symmetric integral operator, and Nowosad [58] has shown that it is positive-definite and that the exact solution φ of (4.6.1) is real. We can therefore use the theory of section 4.2 and set

$$T^*T = K, \qquad \sigma_T = 0, \qquad (4.6.3)$$

$$F(\varphi) = \frac{1}{\varphi}. \qquad (4.6.4)$$

With this $F(\varphi)$ we have $dF/d\varphi = -1/\varphi^2 \leqslant 0$, and hence complementary variational principles can be found. From (4.2.42) and (4.2.46) we have

$$G(T\Psi) \leqslant I(\varphi) \leqslant J(\Phi), \tag{4.6.5}$$

where, by (4.2.38), (4.2.40), and (4.2.41), with $\varphi_0 = 1$ and $W(0, 1) = 0$ to simplify the functionals,

$$J(\Phi) = \int_0^{\pi/2} \{\tfrac{1}{2}\Phi K\Phi - \ln \Phi\} \, dx, \tag{4.6.6}$$

$$I(\varphi) = \int_0^{\pi/2} \{\tfrac{1}{2} - \ln \varphi\} \, dx, \tag{4.6.7}$$

$$G(T\Psi) = \int_0^{\pi/2} \{1 - \tfrac{1}{2}\Psi K\Psi + \ln (K\Psi)\} \, dx. \tag{4.6.8}$$

Here the trial functions Φ and Ψ are intended to be close to the exact solution φ.

Calculations have been performed [4] with the trial functions

$$\Phi = \alpha_1 + \beta_1 x^2, \qquad \Psi = \alpha_2 + \beta_2 x^2, \tag{4.6.9}$$

where the parameters α_1, β_1 and α_2, β_2 were determined by optimizing the functionals J and G. The results are given in Table 4.3. We see from

TABLE 4.3. *Optimum parameters and bounds*

α_1	β_1	J	α_2	β_2	G	$(J - G)/J$
1.36	0.06	0.22364	1.36	0.08	0.22193	0.00764

Table 4.3 that, in terms of the metric $I(\varphi)$, the trial functions (4.6.9) are quite accurate. This accuracy can be improved of course by taking more elaborate trial functions.

4.7. Nonlinear networks

An interesting application of the general theory developed in Chapter 2 can be made to problems of electrical networks, mechanical structures, and problems in economics that can be formulated in terms of network flow. This work is due to Noble [57], and here we shall consider that part dealing with electrical networks. Complementary principles for linear electrical systems have been known since the end of the nineteenth century, but it is only comparatively recently that it has been realized that complementary principles can be obtained for general nonlinear electrical networks [cf. 22, 23, 26, 53].

One of the clues to the existence of complementary variational principles here is that most electrical and mechanical systems involve precisely two different types of variables, for instance voltage and current. The basic equations for such systems can be set up in a formally symmetric way, and to see this we shall look at a particular network problem [57].

Consider the electrical network shown in Fig. 4.1. A branch of the network is a resistor connected to the remainder of the network by precisely two terminals, at nodes. One of the nodes, chosen arbitrarily, is assumed to be at zero potential. Suppose that the other nodes are numbered 1, 2, 3 and that the voltages and input currents at these nodes are denoted by φ_i, α_i ($i = 1$, 2, 3) respectively. The branches are numbered 1–5 and v_j, u_j denote the voltage across and the current in each branch. A direction is assigned (arbitrarily) to each branch, which indicates the positive direction of the current in it, as shown by the arrows on the branches in Fig. 4.1 (a), (b). Relations between the φ_i and v_j can be written down immediately (see Fig. 4.1(c)).

$$
\begin{aligned}
v_1 &= \varphi_1 - \varphi_2 \\
v_2 &= -\varphi_1 \qquad + \varphi_3 \\
v_3 &= \qquad -\varphi_2 + \varphi_3 \\
v_4 &= -\varphi_1 \\
v_5 &= \qquad -\varphi_2.
\end{aligned}
\tag{4.7.1}
$$

The net flow of current at each node must be zero, and this gives relations between the α_i and u_j:

$$
\begin{aligned}
u_1 - u_2 \qquad - u_4 \qquad &= \alpha_1 \\
-u_1 \qquad - u_3 \qquad - u_5 &= \alpha_2 \\
u_2 + u_3 \qquad &= \alpha_3.
\end{aligned}
\tag{4.7.2}
$$

In matrix notation, (4.7.1) and (4.7.2) can be written as

$$
M\varphi = v,
\tag{4.7.3}
$$
$$
M^t u = \alpha,
\tag{4.7.4}
$$

where
$$
M = \begin{pmatrix}
1 & -1 & 0 \\
-1 & 0 & 1 \\
0 & -1 & 1 \\
-1 & 0 & 0 \\
0 & -1 & 0
\end{pmatrix},
\tag{4.7.5}
$$

$$
v^t = (v_1, v_2, v_3, v_4, v_5), \qquad \varphi^t = (\varphi_1, \varphi_2, \varphi_3),
$$
$$
u^t = (u_1, u_2, u_3, u_4, u_5), \qquad \alpha^t = (\alpha_1, \alpha_2, \alpha_3),
\tag{4.7.6}
$$

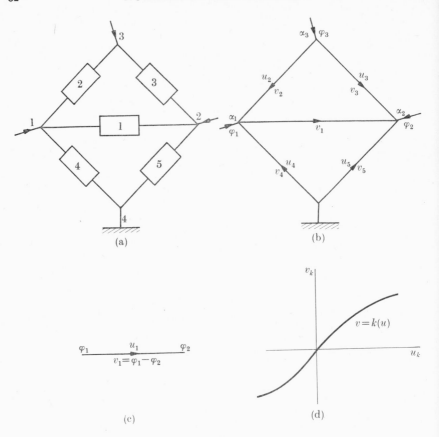

FIG. 4.1. An electrical circuit example

t denoting transpose. It is an important fact that the matrix M^t that occurs in (4.7.4) is the transpose of the matrix M in (4.7.3). This will always be the case for a general network consisting of B branches and N nodes.

In addition to (4.7.3) and (4.7.4) we are given a relation between the current and the voltage in each branch (Fig. 4.1(d)):

$$v_i = k_i(u_i),$$

which we write symbolically as

$$v = k(u). \tag{4.7.7}$$

To make the problem determinate we must be given either α_i or φ_i at each node, or a relation between α_i and φ_i at each node. We shall suppose that α_i is known at each node. Then (4.7.3), (4.7.4), and (4.7.7) give the following pair of equations for the unknown quantities φ, u, the vector α being known:

$$M\varphi = k(u) = \delta W/\delta u, \qquad (4.7.8)$$

$$M^t u = \alpha = \delta W/\delta\varphi. \qquad (4.7.9)$$

These equations correspond to (2.5.21)–(2.5.23) with

$$T = M, \qquad T^* = M^t, \qquad \sigma_T = 0. \qquad (4.7.10)$$

By direct integration, as in (2.5.40), we find that a suitable $W(u, \varphi)$ is given by

$$W(u, \varphi) = \int_0^1 (u - u_0, k\{u_0 + t(u - u_0)\}) \, dt + W(u_0, 0) + \langle \alpha, \varphi \rangle, \qquad (4.7.11)$$

where u_0 is an arbitrary vector, and the inner products are identified with the usual scalar product of matrix multiplication as in (2.1.20) and (2.1.21). The potential or action functional is then given by

$$I(U, \Phi) = (U, M\Phi) - W(U, \Phi), \qquad (4.7.12)$$

$$= \langle M^t U, \Phi \rangle - W(U, \Phi). \qquad (4.7.13)$$

From (4.7.8) and (4.7.9) we see that

$$\frac{\delta^2 W}{\delta u^2} = \frac{dk(u)}{du}, \qquad \frac{\delta^2 W}{\delta\varphi^2} = 0, \qquad (4.7.14)$$

and so, by Theorem 2.5.3, if $(dk/du)(u)$ is positive or negative, complementary variational principles can be obtained. If

$$\frac{dk}{du} \text{ is positive} \qquad (4.7.15)$$

then (2.5.38) and (2.5.39) show that

$$\delta^2 G \leqslant 0 \qquad \text{and} \qquad \delta^2 J \geqslant 0, \qquad (4.7.16)$$

leading to $\qquad\qquad G(U) \leqslant I(u, \varphi) \leqslant J(\Phi), \qquad (4.7.17)$

provided (U, Φ) is close to (u, φ). From the definitions of G and J in (2.5.37) we see that

$$J(\Phi) = (k^{-1}(M\Phi), M\Phi) - W(k^{-1}(M\Phi), \Phi), \qquad (4.7.18)$$

and $\qquad G(U) = -\int_0^1 (U - u_0, k\{u_0 + t(U - u_0)\}) \, dt - W(u_0, 0), \qquad (4.7.19)$

with U subject to the constraint $M^t U = \alpha$. Also

$$I(u,\, \varphi) \;=\; -\int_0^1 (u-u_0,\, k\{u_0+t(u-u_0)\})\; \mathrm{d}t - W(u_0,\, 0). \qquad (4.7.20)$$

An alternative derivation of the complementary principles (4.7.17) is given by Noble [57].

The special case

$$v = k(u) = uR, \qquad (4.7.21)$$

where R is a positive resistance, corresponds to a linear network, and is included in the foregoing theory since $\mathrm{d}k/\mathrm{d}u = R > 0$. In this case, on taking $u_0 = 0$ and $W(0, 0) = 0$ in (4.7.20), we find that

$$I(u,\, \varphi) = -\tfrac{1}{2}(u,\, Ru) = -\tfrac{1}{2}\sum R_i u_i^2, \qquad (4.7.22)$$

which is just the elementary formula for the heat generated in a network.

4.8. Compressible fluid flow

In general the equations of hydrodynamics are highly nonlinear and difficult to solve. It is therefore of some interest to look for associated extremum principles which, if they exist, can be used to provide an approximation method for the construction of fluid flows. To show how the general theory of Chapter 2 can be applied in this connection, we shall consider the steady flow of a nonviscous compressible fluid. For a problem like this the identification of the basic quantity $W(u,\, \varphi)$ is a little less obvious than in some of the earlier applications, and so we shall first develop the canonical equations from the familiar equations of hydrodynamics. Our treatment follows that given by Noble [56].

For the steady flow of a nonviscous compressible fluid in the absence of external forces, the basic equations expressing conservation of mass and momentum are

$$\mathrm{div}(\rho\mathbf{q}) = 0, \qquad (4.8.1)$$

$$(\mathbf{q}.\mathrm{grad})\mathbf{q} = -\frac{1}{\rho}\,\mathrm{grad}\,p, \qquad (4.8.2)$$

where ρ is the density, $\mathbf{q}$ is the velocity, and p is the pressure. In general, p is a function of ρ and a second variable, which we take to be the entropy S. Thus

$$p = g(\rho,\, S). \qquad (4.8.3)$$

Conservation of entropy is expressed by

$$\mathbf{q}.\mathrm{grad}\,S = 0. \qquad (4.8.4)$$

Certain flows are known to be homentropic which means that the entropy S is constant everywhere in space. In this case the pressure p is a function of the density ρ only. The other useful simplification that can occur is that flows are sometimes irrotational, which means

$$\operatorname{curl} \mathbf{q} = 0. \tag{4.8.5}$$

In this case (4.8.2) can be written

$$\tfrac{1}{2} \operatorname{grad} q^2 + \frac{1}{\rho} \operatorname{grad} p = 0,$$

or
$$\tfrac{1}{2} q^2 + \int \frac{\mathrm{d}p}{\rho} = \text{constant along a streamline}, \tag{4.8.6}$$

where $q^2 = \mathbf{q} \cdot \mathbf{q}$. Since p is a function of ρ, (4.8.6) can be written as

$$f'(\rho) - \tfrac{1}{2} q^2 = 0, \tag{4.8.7}$$

where we have absorbed the constant into $f'(\rho)$. We see that

$$\frac{\mathrm{d}p}{\mathrm{d}\rho} = -\rho f''(\rho) = c^2, \tag{4.8.8}$$

where c is the speed of sound in the fluid. Integration of (4.8.8) leads to

$$p = f(\rho) - \rho f'(\rho)$$
$$= f(\rho) - \tfrac{1}{2} \rho q^2. \tag{4.8.9}$$

Bateman [20] in his work on extremum principles considered the case of homentropic, irrotational flow in two dimensions. This is an important case historically, and will serve to illustrate the use of complementary variational principles. In two dimensions $\mathbf{q} = (q_x, q_y)$ and (4.8.5) becomes

$$\frac{\partial q_x}{\partial y} - \frac{\partial q_y}{\partial x} = 0, \tag{4.8.10}$$

which can be satisfied by introducing the velocity potential φ such that

$$\frac{\partial \varphi}{\partial x} = q_x, \qquad \frac{\partial \varphi}{\partial y} = q_y, \tag{4.8.11}$$

or
$$\operatorname{grad} \varphi = \mathbf{q}. \tag{4.8.12}$$

We now have the basic equations in (4.8.1), (4.8.7), and (4.8.12). These can be rewritten as

$$\text{grad } \varphi = \frac{1}{\rho}\mathbf{u} = \frac{\partial H}{\partial \mathbf{u}}, \tag{4.8.13}$$

$$-\text{div }\mathbf{u} = 0 = \frac{\partial H}{\partial \varphi}, \tag{4.8.14}$$

$$\rho^2 f'(\rho) - \tfrac{1}{2}u^2 = 0, \tag{4.8.15}$$

where
$$\mathbf{u} = \rho\mathbf{q}. \tag{4.8.16}$$

Since ρ is a function of $(u_x^2 + u_y^2)^{\frac{1}{2}} = u$ only, by (4.8.15), it follows from (4.8.13) and (4.8.14) that the generalized Hamiltonian H is a function of u only. To proceed, it is useful to introduce

$$H(u(\rho)) = \Omega(\rho) \text{ say.} \tag{4.8.17}$$

Then we can readily show that

$$\Omega(\rho) = f(\rho) + \rho f'(\rho) = f(\rho) + \tfrac{1}{2}\rho q^2, \tag{4.8.18}$$

by using (4.8.13) and (4.8.15). This is the basic Hamiltonian which appears in the potential $I(u, \varphi)$ of the associated variational problem. Assuming homogeneous boundary conditions, we therefore find from (2.3.10) and (2.3.11) that

$$I(\mathbf{u}, \varphi) = \int \{\mathbf{u}.\text{grad }\varphi - f(\rho) - \tfrac{1}{2}\rho q^2\}\, \mathrm{d}V, \tag{4.8.19}$$

$$= \int \{(-\text{div }\mathbf{u})\varphi - f(\rho) - \tfrac{1}{2}\rho q^2\}\, \mathrm{d}V. \tag{4.8.20}$$

The functional $J(\varphi_t)$, where φ_t denotes a trial function, is given by

$$J(\varphi_t) = I(\mathbf{u}(\varphi_t), \varphi_t), \tag{4.8.21}$$

where
$$\mathbf{u}(\varphi_t) = \rho_t\,\text{grad }\varphi_t = \rho_t\mathbf{q}_t \tag{4.8.22}$$

is the solution of (4.8.13). Hence

$$J(\varphi_t) = \int \{\rho_t(\text{grad }\varphi_t)^2 - f(\rho_t) - \tfrac{1}{2}\rho_t q_t^2\}\, \mathrm{d}V$$

$$= \int \{\tfrac{1}{2}\rho_t q_t^2 - f(\rho_t)\}\, \mathrm{d}V$$

$$= -\int p(\rho_t)\, \mathrm{d}V, \tag{4.8.23}$$

by (4.8.9). Here ρ_t is a function of the trial function φ_t.

The functional $G(\mathbf{u}_t)$, where $\mathbf{u}_t$ denotes a trial function, is given by

$$G(\mathbf{u}_t) = I(\mathbf{u}_t, \varphi(\mathbf{u}_t)), \tag{4.8.24}$$

where $\varphi(\mathbf{u}_t)$ is the solution of (4.8.14), which in this case actually imposes the constraint

$$\operatorname{div} \mathbf{u}_t = 0 \qquad (4.8.25)$$

on the trial $\mathbf{u}_t$. Hence from (4.8.20)

$$G(\mathbf{u}_t) = -\int \{f(\rho_t) + \tfrac{1}{2}\rho_t q_t^2\}\, \mathrm{d}V$$

$$= -\int \{p(\rho_t) + \rho_t q_t^2\}\, \mathrm{d}V \qquad (4.8.26)$$

by (4.8.9). Here the implication is that ρ_t and q_t are functions of the trial function $\mathbf{u}_t$, which is subject to the constraint $\operatorname{div} \mathbf{u}_t = 0$ in V.

Equations (4.8.23) and (4.8.26) contain the p and $p + \rho q^2$ functions of Bateman [20].

To examine the extremum property of $J(\varphi_t)$ we consider variations of q_t about q and expand

$$J(\varphi_t) = J(\varphi) - \tfrac{1}{2}\epsilon^2 \int \sum_{i,j} \delta q_i \frac{\partial^2 p}{\partial q_i \cdot \partial q_j}\, \delta q_j\, \mathrm{d}V + \ldots, \qquad (4.8.27)$$

where $\epsilon\, \delta q_i = q_{ti} - q_i$. From the discussion of Sewell [74] we see that uniqueness of solution requires that $-p$ be convex. This is ensured if

$$\sum_{i,j} \delta q_i \frac{\partial^2 p}{\partial q_i\, \delta q_j}\, \partial q_j \leqslant 0. \qquad (4.8.28)$$

Now by (4.8.9)

$$\frac{\partial^2 p}{\partial q_i \cdot \partial q_j} = \rho\left(\frac{q_i q_j}{c^2} - \delta_{ij}\right), \qquad (4.8.29)$$

and so (4.8.28) is satisfied by subsonic flow. In this case we have the minimum principle

$$J(\varphi) < J(\varphi_t) \qquad (4.8.30)$$

for φ_t sufficiently close to the exact solution φ.

Under the same conditions of subsonic flow one can similarly show that the complementary maximum principle

$$G(\mathbf{u}_t) < G(\mathbf{u}) = J(\varphi) \qquad (4.8.31)$$

holds for $\mathbf{u}_t$ sufficiently close to the exact solution $\mathbf{u}$.

For an alternative derivation of these extremum principles in a more general context, we refer to the work of Sewell [74]. A discussion of variational principles when dissipative forces are present can be found in the book by Yourgrau and Mandlestam [84].

SUMMARY

This chapter has been concerned with applications of the theory of Chapter 2 to a class of nonlinear boundary-value problems. The associated complementary variational principles were illustrated by various examples taken from mathematical physics. These examples were in (1) hydrodynamics, involving the Liouville and Poisson–Boltzmann equations and the equations of compressible fluid flow, (2) the Thomas–Fermi statistical theory of atomic structure, (3) information theory, dealing with a certain nonlinear integral equation, and (4) the theory of nonlinear electrical networks.

This book is intended to provide an introduction to the theory and applications of complementary variational principles. The treatment was based on local variational theory, leaving several aspects of the subject, such as global theory and numerical topics in the applications, to be developed elsewhere.

REFERENCES

1. ALEXANDER, A. E. and JOHNSON P. *Colloid science.* Oxford University Press (1949).
2. ANDERSON N. and ARTHURS, A. M. *J. math. Phys.* **9**, 2037 (1968).
3. ——, ——, *Nuovo Cim. lett.* **1**, 119 (1969).
4. ——, ——, *Nuovo Cim. lett.* **2**, 631 (1969).
5. ——, ——, and ROBINSON, P. D. *Nuovo Cim.* **57B**, 523 (1968).
6. ——, ——, ——, *J. math. Phys.* **10**, 1498 (1969).
7. ——, ——, ——, *J. Inst. Math. Appl.* **5**, 422 (1969).
8. ——, ——, ——, to *J. Phys. A.* **3**, 1 (1970).
9. ANSELONE, P. M. and RALL, L. B. *Univ. Wis. Math. Res. Center, Report No.* 492 (1964).
10. ARTHURS, A. M. *Proc. R. Soc.* **A298**, 97 (1967).
11. ——, *Phys. Rev.* **176**, 1730 (1968).
12. ——, *Proc. Camb. phil. Soc. math. phys. Sci.* **65**, 803 (1969).
13. ——, *Proc. Camb. phil. Soc. math. phys. Sci.* **66**, 399 (1969).
14. ——, *J. Math. Anal. Appl.* (1970).
15. ——, and ROBINSON, P. D. *Proc. R. Soc.* **A303**, 497 (1968).
16. ——, ——, *Proc. R. Soc.* **A303**, 503 (1968).
17. ——, ——, *Proc. Camb. phil. Soc. math. phys. Sci.* **65**, 535 (1969).
18. ——, ——, *Proc. Camb. phil. Soc. math. phys. Sci.* **66**, 433 (1969).
19. AUER, P. L. and GARDNER, C. S. *J. chem. Phys.* **23**, 1545 (1955).
20. BATEMAN, H. *Proc. natn. Acad. Sci. U.S.A.* **16**, 816 (1930).
21. ——, *Partial differential equations.* Cambridge University Press (1959).
22. BIRKHOFF, G. *Q. appl. Math.* **21**, 160 (1963).
23. —— and DIAZ, J. B. *Q. appl. Math.* **13**, 431 (1956).
24. BUSH, V. and CALDWELL, S. H. *Phys. Rev.* **38**, 1898 (1931).
25. CASE, K. M., DE HOFFMAN, F., and PLACZEK, G. *Introduction to the theory of neutron diffusion,* Vol. 1. Los Alamos Scientific Laboratory (1953).
26. CHERRY, C. *Phil. Mag.* (7) **42**, 1161 (1951).
27. CONLAN, J., DIAZ, J. B., and PARR, W. E. *J. math. Phys.* **2**, 259 (1961).
28. COURANT, R. and HILBERT, D. *Methods of mathematical physics,* Vol. 1. Interscience Publishers, New York (1953).
29. DALGARNO, A. In *Quantum theory* (ed. D. R. BATES) Vol. 1. ch. 5. Academic Press, New York (1961).
30. DAVIS, H. T. *Introduction to nonlinear differential and integral equations.* Dover Publications, New York (1962).
31. DAVISON, B. *Neutron transport theory.* Oxford University Press (1957).

32. DRESNER, L. *J. math. Phys.* **2**, 829 (1961).

33. FEYNMAN, R. P., LEIGHTON, R. B., and SANDS, M. *The Feynman lectures on physics*, Vol. 2. Addison-Wesley, Reading, Mass. (1964).

34. FRIEDMAN, B. *Principles and techniques of applied mathematics*. Wiley, New York (1956).

35. GELFAND, I. M. and FOMIN S. V. *Calculus of variations*. Prentice-Hall, Englewood Cliffs, New Jersey (1963).

36. GOULD, S. H. *Variational methods for eigenvalue problems*, 2nd edition. University of Toronto Press (1966).

37. HAHN, Y., O'MALLEY, T. F., and SPRUCH, L. *Phys. Rev.* **128**, 932 (1962); **130**, 381 (1963).

38. HILLE, E. and PHILLIPS, R. S. *Functional analysis and semi-groups*. *Amer. math. Soc. Colloq. Publ.* Vol. 31, Providence (1957).

39. HIRSCHFELDER, J. O., BYERS BROWN, W., and EPSTEIN, S. T. *Adv. Quantum Chem.* **1**, 255 (1964).

40. HYLLERAAS, E. A. *Z. Phys.* **65**, 209 (1930).

41. JOHNSON, M. W. *Univ. Wis. Math. Res. Center Report* No. 208 (1960).

42. KATO, T. *Math. Annln.* **126**, 253 (1953).

43. KIRKWOOD, J. G. and RISEMAN, J. *J. chem. Phys.* **16**, 565 (1948).

44. KOBAYASHI, S., MATSUKUMA, T., NAGAI, S., and UMEDA, K. *J. phys. Soc. Japan* **10**, 759 (1955).

45. KOMKOV, V. *J. Math. Anal. Appl.* **14**, 511 (1966).

46. LANCZOS, C. *The variational principles of mechanics*, 3rd edition. University of Toronto Press (1966).

47. LANDAU, L. D. and LIFSHITZ, E. M. *Quantum mechanics*. Pergamon Press, Oxford (1958).

48. LEVINE, H. and SCHWINGER, J. *Phys. Rev.* **74**, 958 (1948); **75**, 1423 (1949).

49. LLEWELLYN-JONES, F. *The glow discharge*. Methuen, London (1966).

50. LONGMIRE, C. L. *Elementary plasma physics*, ch. 8. Interscience Publishers, New York (1963).

51. MARK, J. C. *Natn. Res. Coun. of Canada Report* CRT-338, Montreal (1945).

52. MIKHLIN, S. G. *Variational methods in mathematical physics*. Pergamon Press, Oxford (1964).

53. MILLAR, W. *Phil. Mag.* (7) **42**, 1150 (1951).

54. MOISEIWITSCH, B. L. *Variational principles*. Interscience Publishers, New York (1966).

55. NOBLE, B. *Univ. Wis. Math. Res. Center Report No.* 473 (1964).

56. ——, *Variational principles in compressible fluid flow*. Unpublished lecture notes, University of Wisconsin (1965).

57. ——, *Univ. Wis. Math. Res. Center Report No.* 643 (1966).

58. NOWOSAD, P. *J. Math. Anal. Appl.* **14**, 484 (1966).

59. PARS, L. A. *Calculus of variations*. Heinemann, London (1962).

REFERENCES   91

60. PÓLYA, G. and SZEGÖ, G. *Isoperimetric inequalities in mathematical physics*. Princeton University Press (1951).
61. POMRANING, G. C. *J. math. Phys.* **8**, 2096 (1967).
62. ——, *J. Math. Phys.* **47**, 155 (1968).
63. ——, and LATHROP, K. D. *Nucl. Sci. Engng.* **29**, 305 (1967).
64. PRAGER, S. and HIRSCHFELDER, J. O. *J. chem. Phys.* **39**, 3289 (1963).
65. RALL, L. B. In *Nonlinear integral equations* (ed. P. M. ANSELONE) pp. 155–189. University of Wisconsin Press, Madison (1964).
66. ——, *J. Math. Anal. Appl.* **14**, 174 (1966).
67. RAYLEIGH, LORD. *Phil. Mag.* **47**, 566 (1899).
68. RITZ, W. *J. reine angew. Math.* **135**, 1 (1908).
69. ROBERTS, R. E. *Phys. Rev.* **170**, 8 (1968).
70. ROBINSON, P. D. *J. Phys.* **A. 2**, 193 (1969).
71. ——, and ARTHURS, A. M. *J. math. Phys.* **9**, 1364 (1968).
72. SALTZBERG, B. R. and KURZ, L. *Bell Syst. tech. J.* **44**, 235 (1965).
73. SCHRADER, D. M. *J. math. Phys.* **8**, 870 (1967).
74. SEWELL, M. J. *J. Math. Mech.* **12**, 495 (1963).
75. SHAMPINE, L. F. *Num. Math.* **12**, 410 (1968).
76. SPRUCH, L. and ROSENBERG, L. *Phys. Rev.* **116**, 1034 (1959).
77. STAKGOLD, I. *Boundary value problems of mathematical physics*, Vol. 2. Macmillan, New York (1968).
78. SYNGE, J. L. *The hypercircle in mathematical physics*. Cambridge University Press (1957).
79. TEMPLE, G. *Proc. R. Soc.* **A119**, 276 (1928).
80. TREFFTZ, E. *Math. Annln.* **100**, 503 (1928).
81. ULLMAN, N. and ULLMAN, R. *J. math. Phys.* **7**, 1743 (1968).
82. VAINBERG, M. M. *Variational methods for the study of nonlinear operator equations*. Holden-Day, San Francisco (1963).
83. WEINBERG, A. M. and WIGNER, E. P. *The physical theory of neutron chain reactors*. University of Chicago Press (1958).
84. YOURGRAU, W. and MANDLESTAM, S. *Variational principles in dynamics and quantum theory*, 3rd edition. Pitman, London (1968).

SUBJECT INDEX